Dying Green

Critical Issues in Health and Medicine

Edited by Rima D. Apple, University of Wisconsin–Madison;
Janet Golden, Rutgers University–Camden; and
Rana A. Hogarth, University of Illinois at Urbana–Champaign

Growing criticism of the U.S. healthcare system is coming from consumers, politicians, the media, activists, and healthcare professionals. Critical Issues in Health and Medicine is a collection of books that explores these contemporary dilemmas from a variety of perspectives, among them political, legal, historical, sociological, and comparative, and with attention to crucial dimensions such as race, gender, ethnicity, sexuality, and culture.

For a list of titles in the series, see the last page of the book.

Dying Green

A Journey through End-of-Life Medicine in Search of Sustainable Health Care

CHRISTINE VATOVEC

Rutgers University Press
New Brunswick, Camden, and Newark, New Jersey
London and Oxford

Rutgers University Press is a department of Rutgers, The State University of New Jersey, one of the leading public research universities in the nation. By publishing worldwide, it furthers the University's mission of dedication to excellence in teaching, scholarship, research, and clinical care.

Library of Congress Cataloging-in-Publication Data

Names: Vatovec, Christine, author.
Title: Dying green : a journey through end-of-life medicine in search of sustainable health care / Christine Vatovec.
Description: New Brunswick : Rutgers University Press, [2023] | Series: Critical issues in health and medicine | Includes bibliographical references and index.
Identifiers: LCCN 2022037002 | ISBN 9781978832107 (paperback) | ISBN 9781978832114 (hardback) | ISBN 9781978832121 (epub) | ISBN 9781978832145 (pdf)
Subjects: LCSH: Medical care—Environmental aspects. | Medical economics. | Terminal care—Economic aspects. | Medical wastes—Environmental aspects. | Sustainability. | BISAC: HEALTH & FITNESS / Health Care Issues | SOCIAL SCIENCE / Death & Dying
Classification: LCC RA394 .V38 2023 | DDC 338.4/73621—dc23/eng/20221202
LC record available at https://lccn.loc.gov/2022037002

A British Cataloging-in-Publication record for this book is available from the British Library.

rutgersuniversitypress.org

For my father, Edward Rudy Vatovec

Contents

	Introduction	1
1	Focal Point: End-of-Life Medical Care	11
2	Medical Waste	27
3	Medical Supplies	55
4	Pharmaceuticals	77
5	Patients	99
6	Conclusions and Practical Implications	111
	Acknowledgments	121
	Appendix A A Note on Methods	123
	Appendix B A Note on Theory	127
	Appendix C Institutional Data on Materials Used at Hopewell Hospital and Baluster Hospice	135
	Notes	139
	References	151
	Index	165

Contents

Dying Green

Introduction

Two intertwined stories occupy the center of this book. One story asks us to consider the environmental sustainability of health care, and, in particular, how clinical practices and medical decision making drive the unintended consequences of medicine. The other story suggests that by focusing our sustainability analysis on the clinical care provided to terminally ill cancer patients, we can identify points of intervention that could lead to better outcomes for both patients and the planet. Together, these stories derive from the 255 hours I spent as an ethnographer observing and interviewing health care staff as they provided end-of-life care to cancer patients in three distinct medical settings. As these stories unfold through the following pages, I share insights on the many opportunities for reducing the environmental impacts of medical practices in general, while also enhancing care for the dying in particular. Such an analysis is both useful and necessary when we stop to consider whether we want our health care system to support the well-being of people both within and beyond the walls of our medical facilities.

Environmental Sustainability and Health Care

Every day, thousands of people walk through the door of their local hospital in search of medical care. This simple action relates to environmental sustainability in two ways. First is the ecological footprint of the facility.

Hospitals require large volumes of natural resources—land, water, wood, metal, energy—to provide heating, cooling, lighting, food, laundry, disinfection, waste disposal, parking, and other services twenty-four hours a day, 365 days each year.[1] The size of these facilities ranges from small towns of a few hundred staff members, to small cities that support thousands of employees, patients, and visitors each day. The harvesting, manufacturing, transporting, and disposal of materials used to provide health care services in the United States, and the energy required to support each of these processes, result in tremendous cumulative effects on the environment, workers, and public health.

Second is the ecological footprint of providing medical care itself. The assessment, diagnosis, and treatment of an individual require a vast array of technological devices, medical supplies, and pharmaceuticals, each of which has an ecological story of its own. Every resource used to deliver medical care—be it intravenous tubing, the medication flowing through the tube, or the electricity powering the IV pump that controls the drug's flow—has unintended impacts on human health and the environment, either directly or indirectly. Ecosystems that provide the raw materials of medical supply chains are degraded through harvesting and extraction processes, affecting biodiversity. Toxins released throughout the life cycle of medical supplies lead to acute and chronic health effects among patients, workers, and nearby communities. Pharmaceuticals that enter the environment through manufacturing, excretion, and disposal pollute aquatic ecosystems and contaminate drinking water. Carbon emissions from the factories, trucks, and cargo ships that produce and transport medical materials contribute to the destabilization of the global climate. Each of these challenges represents an unintended consequence of health care, and each is driven by clinical practices and medical decision making.

These negative outcomes of medical care raise both the ethical dilemma that to improve the health of patients we may unintentionally harm the health of the workers and communities on which the health care system depends, and the irony that health care contributes to ecosystem degradation, biodiversity loss, and climate instability, which negatively affect human health on a global scale.

One of the clearest examples of the challenge that medical care poses to environmental sustainability is its impact on climate change. In September 2021, over two hundred international health journals concurrently published an editorial calling for action to reduce greenhouse gas emissions and calling climate change the greatest public health threat of our time.[2] Recent estimates show that globally, the health care sector accounts for up to 5 percent of greenhouse gas emissions. In the United States this statistic rises to between 8.5 and 10 percent—the highest of all industrialized nations. If the U.S. health care sector were itself a country, it would rank thirteenth for global greenhouse gas emissions. These emissions contribute to the destabilization of earth's climate, which in turn affects human health in myriad ways. The National Climate and Health Assessment indicates seven primary health impacts of climate change. These include temperature-related illness and mortality, respiratory and allergenic effects related to air quality, injury and death from extreme events, acute and chronic health effects of vector-borne disease, illness resulting from poor water quality, risks posed by food insecurity, and the cumulative burden climate change poses to mental health and well-being.[3] The health care sector's contributions to the burden of climate-related illness result from the vast energy demands of the infrastructure needed to provide patient care.

Some might argue that these externalities of health care are unfortunate but necessary in order to promote the health of our citizens. However, despite having the most expensive health care system in the world, the United States lags far behind other countries in several health outcomes. In 2018, the United States allocated nearly 17 percent of gross domestic product to health care, representing over $10,000 per capita on health spending.[4] As such, the United States held the top rank of all countries for health care expenditures, with nearly double the average expense of the thirty-six high-income member countries of the Organization for Economic Cooperation and Development (OECD). Despite the amount of money we allocate to health care, Americans have the highest rate of obesity, the highest rate of hospitalization and premature death from preventable diseases, and the highest disease burden among adults when compared to our peer countries. Perhaps most tellingly, despite our higher health care expenditures, Americans have the shortest life expectancy at birth. We spend 31 percent more on health care per capita in the United States than Switzerland, yet the Swiss live on average four years longer than Americans. The comparison with Japan is even starker, with our health care spending 55 percent higher, and life expectancy five years lower. The high cost of care in the United States is not achieving the health benefits that we might expect.

To move toward a sustainable health care system that improves human health while also respecting planetary boundaries, we must identify points of intervention within the health care system that will allow us to implement practices and policies in support of both high-quality patient care and a flourishing natural environment.

The History of Health Care Greening

This book is not the first time that the environmental sustainability of health care has been under question. A series of developments in the history of health care greening provide insight into the successes and limits of current efforts, each of which provides evidence of the desire and capacity for change within the system.

First, in 1996, the U.S. Environmental Protection Agency released a report on medical waste incinerators that identified these facilities as both the leading cumulative source of dioxin emissions in the country and a significant source of mercury pollution. In response, a coalition of twenty-eight health care organizations in California joined together to form the nonprofit organization Health Care Without Harm (HCWH) with a mission to "transform health care worldwide so that it reduces its environmental footprint, becomes a community anchor for sustainability and a leader in the global movement for environmental health and justice." Today, HCWH is an international leader in the movement toward sustainable health care with thousands of hospitals and health partners participating in its work in more than fifty countries. The primary outcomes and successes of HCWH to date include increased advocacy for and use of renewable energy in health care, the phase-out of medical waste incineration, the elimination of the global production of mercury-based thermometers, the transition toward environmentally preferred medical supplies and sustainable food systems, and the increased use of green building technologies in health care facilities.[5]

Second, in 2001, HCWH collaborated with the Environmental Protection Agency, the American Hospitals Association, and the American Nurses Association to form what has evolved into the nation's leading nonprofit membership organization dedicated to environmental sustainability in health care: Practice Greenhealth.[6] This new endeavor provided a place for health care institutions to access educational resources, technical expertise, and performance measurement tools to help reduce health care waste and minimize toxins used in medical supplies and facilities. One of Practice Greenhealth's premier resources is the Greenhealth Academy, which offers a variety of training opportunities to help health care institutions increase recycling efforts, transition to green cleaning chemicals, and identify opportunities for energy use reduction through modeling. They also offer resources to assist hospitals with procurement of environmentally preferred medical supplies through group purchasing organizations that help reduce the impact of medical supply chains on human health and the environment.

Third, in 2007, the Institute of Medicine of the National Academies released the report *Green Healthcare Institutions: Health, Environment, and Economics*, which was the result of a workshop on this topic that had been convened by the Roundtable on Environmental Health Sciences, Research, and Medicine.[7] Although the introduction to the report defined green health care as "the incorporation of environmentally friendly practices into healthcare delivery," the workshop and report focused on, "the environmental and health impacts related to the design, construction, and operation of healthcare facilities," and specifically green building. The report called for the health care industry to provide leadership in environmental responsibility and highlighted a series of best practices for implementing sustainability principles in facility design and operation.

Most recently, Yale University's Center on Climate Change and Health has hosted a series of events pulling together academic sustainability experts, clinicians, health care facility architects and operations managers, and health care administrators to address sustainability challenges within the industry.[8] Among these events were the 2018 "Workshop on Environmental Sustainability in Clinical Care," and 2021 "Care without Carbon: The Road to Sustainability in U.S. Health Care." Together, these events call for systematic change across the health care sector and provide frameworks and models for implementing best practices for pollution prevention and carbon emission reductions.

These efforts have driven significant improvements in the ecological footprint of health care, reduced costs while increasing patient satisfaction, and provided a foundation for further positive change within the industry.[9] Despite these successes, the focus of these efforts has been limited to changes in facilities, operational management, and the procurement of environmentally preferred materials and supplies. To truly move toward the goal of sustainable health care, we must broaden our focus to include medical decision making and clinical practices themselves—those practices that drive the ecological impacts of health care.

This book builds on the early successes of the health care greening movement by offering a comprehensive view from inside the medical system on the environmental sustainability of care provided at the patient bedside. In essence, this investigation focuses on two questions: (1) What unintended environmental and public health consequences arise from clinical practices and medical decision making, and (2) What are the social factors that govern these outcomes? By examining the environmental flows of medical supplies, pharmaceuticals, and waste that result from clinical care, along with the political economy, institutional forces, and cultural factors that guide those materials, this analysis reveals several opportunities for further increasing environmental sustainability within health care.

The Ethics of Health Care Greening

At first glance, the question of environmental sustainability within health care may seem to place the health of patients and the health of the environment in opposition. Some may argue that we can either offer patients the best possible care, or we can take account of the environmental impacts of our health care system, but we cannot do both. A lively debate over the ethical dilemma of instilling environmental concerns into medical decision making began in the late 1990s and offers important perspectives that we must consider when determining how best to move forward with envisioning a more sustainable health care system that is supportive of both patients and the planet.[10]

On one side of the debate, Andrew Jameton and Jessica Pierce have argued that health care must remain within sustainable limits of resource use to ensure both human and planetary health. For example, in the case of latex medical gloves Jameton asked, "Suppose that the process of obtaining latex gloves is part of the gradual erosion of the Malaysian environment, and that workers in latex factories are poorly paid. Now, should or would you refuse to use latex gloves? Should or would you even be more selective in their use?" This argument suggests that by incorporating environmental costs into health care options, patients attain true autonomy by being able to make decisions based on the "fullest possible awareness" of their implications.

On the other side of the debate, Paul Carrick warned, "We should be cautious of importing global environmental ethical theories into our health care ethics precisely because these environmental theories, often with the best intentions, may undermine respect for individual human life." He continued, "When an environmental theory pressures the terminally ill to quit life in favor of the global claims of ecological paternalism, then the dignity, autonomy, and inherent value of that dying patient's life are diminished."

Both perspectives are valid and require consideration as we grapple with identifying a path forward that encompasses patient care directly within efforts toward health care sustainability. I was fortunate to encounter this debate early in my work—after identifying end-of-life care as the focus of my research but before beginning data collection—for I often considered these arguments as I observed patient care. In retrospect, it seems serendipitous that my work centered on end-of-life health care because this setting forces us to face challenging subjects directly. In the pages that follow, my analysis provides evidence that I believe helps bring resolution to the opposing arguments of incorporating environmental concerns into medical decision making. My findings suggest that by promoting practices that encourage the dignity, autonomy, and inherent value of dying patients, we may actually create the circumstances under which a more ecologically sustainable health care system can emerge. So rather than leading the way, an environmental ethic may actually follow as a natural outcome from acknowledging and better valuing the unique human experience of dying.

Chapter Overview

Death and the medical models that support the dying are the focus of chapter 1. The chapter introduces end-of-life medical care as a unique setting for investigating the environmental and public health consequences of health care, and the social factors that govern these impacts. The chapter offers three key reasons for focusing our investigation specifically on end-of-life care for terminally ill cancer patients: (1) end-of-life cancer care is a high-cost setting, which suggests high volumes of resource use, (2) it serves a large number of patients each year, suggesting large cumulative impacts, and (3) the three distinct end-of-life cancer care medical models that exist allow for comparative analysis. The chapter provides a brief overview of the history of end-of-life care in the United States and introduces the three health care settings in which I conducted my observations. Chapter 1 shares my initial insights into social factors that drive clinical care and, in particular, their different philosophies of care.

Next, my analysis turns toward the ecological impacts incurred within each of my three research settings, and the institutional, cultural, and economic factors that drove these outcomes. Chapter 2 provides an overview of the five waste streams that flowed from my inpatient research settings, describes their ultimate fate, and provides insights into the social and economic factors associated with waste generation along with points of intervention for reducing their unintended impacts. The chapter compares the volumes of waste generated by each of my three research settings. This analysis begins to show how the differences in goals and philosophies that guide clinical decision making across the three inpatient units lead to disparate ecological impacts.

Likewise, chapter 3 shows how factors that govern clinical care lead to differences in the volume of medical supplies used across the three inpatient units, and therefore differences in the cumulative impacts that medical supply chains have on ecological and human communities. The chapter presents the pioneering "ethical life cycle analysis" of Jessica Pierce and Christina Kerby as a framework for understanding the cumulative impacts of medical supply chains on the environment, workers, and public health, and builds on their work to examine the impacts of nitrile medical gloves and polyvinyl chloride plastics. Chapter 3 provides insights on the role that institutional culture plays in determining the volume of supplies used within clinical settings.

Chapter 4 focuses on the unintended consequences of pharmaceuticals. The discussion focuses on the hazards of occupational exposures to chemotherapy and the environmental impacts that arise throughout pharmaceutical supply chains. The chapter describes how the clinical practices of overprescribing and overdispensing lead to a large volume of unused medications that require disposal. It also explores how concerns about the conflicting risks of drug diversion and aquatic pollution govern the fate of unused medications and lead to pharmaceutical pollution entering the environment.

Chapter 5 shares my insights on what appears to be a primary driver of end-of-life health care decision making: communication between patients and providers. In particular, the chapter presents evidence from interviews, observations, and the academic literature that suggest having goal-setting conversations with patients early within their care, and revisiting these conversations as disease progresses, allows patients the choice to transition away from conventional care and toward hospice care when medically appropriate. Without these conversations, patients are more likely to continue conventional life-prolonging treatments beyond the time when they are medically beneficial, at the cost of quality of life and in some cases longevity. The chapter also provides an overview of the reported benefits that patients and families receive from the early transition to hospice and its relationship to environmental sustainability.

Chapter 6 opens with my observations of a death scene within the hospice facility that stood in stark contrast to the practice of death at the hospital. This scene highlights the very material ways in which the different philosophies of care in these medical models guide the patient and family experience of death and dying, and drive their varied environmental impacts. Finally, the analysis turns to summarize and compare the cumulative ecological impacts that resulted from the three medical models under investigation. Each setting had environmental impacts that we must address if we wish to move toward a more sustainable health care system, and each offered insight into ways to overcome these challenges.

In the end, this book does not offer a clear and concise answer to the question of how best to provide high-quality medical care that fully accounts for its impacts on the environment and public health. Rather, the analysis raises questions for us to grapple with together as a global, ecological citizenry in order to envision a health care system that aligns medical practices with patient preferences and ecological boundaries.

1

Focal Point

End-of-Life Medical Care

The warm air of the bus exhaled me with a rush onto the cold midwinter curb less than a block from the hospital. As the bus pulled away into the gritty build-up of dirt on the snow-covered street, I held tightly to a friend's handwritten directions to the hospital's hidden back entrance. He had told me it was the quickest way to enter the warmth of the building after a wintery walk from the bus stop.

This day marked the beginning of nearly two years of observing and interviewing health care staff—from physicians, nurses, and pharmacists to environmental services workers—to gain insight into the unintended consequences of medical care on the natural environment and public health. Following my friend's directions, I made my way up a short hill alongside a staff parking lot, where an entrance gate clicked open and closed with the arrival of the morning shift. As I neared the building, I sensed the heavy smell of tobacco smoke and saw the stubs of cigarettes littering the ground near the outdoor ashtray. Backed up to the door sat a black, nondescript minivan, blocking my path. Nervous that I would be late on my first day, I squeezed through the narrow opening between the van and the edge of the building and was startled by two men in suits and ties. They were pushing a gray metal gurney that held a zippered black body bag. I halted and made way for the caretakers. The men greeted me tentatively and carried on with their task. The wheeled metal legs of the gurney clanged as they folded into the van, the black bag and its contents rocking in response. I walked a bit more quickly for the final few steps to the door and pushed the silver handle down to enter the building. This moment marked my first of many encounters with the practice of death and dying at Hopewell Hospital.[1]

A few months later, a man died early one Friday morning on the palliative care unit. When I arrived at the start of the day shift, the patient's family was already gone, the nurses had removed the patient's IV line and catheter, and the unit clerk had called both the morgue and the family's preferred funeral home to let them know the body would soon be on its way. I watched as the nursing staff prepared to transport the body down to the morgue. The nurses usually preferred to use a special gurney modified to look somewhat like a box on wheels—a good way to disguise a dead body from the eyes of passersby—but this morning there were no morgue gurneys available. Instead, the nurses moved the body from the bed to a stretcher, put the side rails up, and draped a white sheet over the top. The nose, belly, and toes of the man were still discernible from beneath the sheet, and the nurses were concerned that anyone they passed in the hallway would know it was a body so they did everything they could to make their cargo less apparent. They pulled the sheet taut across the top of the side rails and taped it to the sides of the stretcher. One nurse gently reached beneath the sheet and turned the man's head to the side so his nose would no longer poke up from beneath the drape. Another nurse placed a pillow over the man's feet so his toes became invisible. When they had done all they could in the time available, we headed off down the hallway to the freight elevator that would lead us to the morgue, where in turn the funeral home director would collect and remove the body through the back door.

These two experiences framed my understanding of the hospital's cultural approach to death and provided insight into the social factors that govern medical decision making and associated resource use that I share throughout the following pages. At the same time, these experiences supported my assumption that end-of-life care would provide an excellent focus for investigating the broader ecological impacts of clinical decisions.

End-of-Life Care as a Focal Point for Understanding the Unintended Consequences of Medical Decision Making

Why focus on end-of-life care among cancer patients to investigate the factors that govern unintended environmental and public health consequences of medical care? My journey to this place had begun over a decade earlier when my father was diagnosed with leiomyosarcoma, a type of soft tissue cancer that was wreaking havoc throughout his abdominal organs. At the time, I was an undergraduate student at Cornell University studying ecology, conservation biology, and natural resources, and I viewed my father's treatment through an ecological lens. As an ecologically minded organic dairy farmer, he prompted me with questions about what effect his chemotherapy treatment would have on the environment. After he died, I went on to earn a master's degree in conservation biology and worked on a number of ecological research and restoration projects across the country. But my thoughts often returned to the questions my father had posed about the ecological connections between medicine, the environment, and health.

Years later, when I approached my PhD adviser, an environmental sociologist, with these questions, he greeted my inquiry with sincere interest and challenged me to first determine what work had already been done on this topic. As I described in the introduction to this book, many people had already begun questioning the environmental sustainability of the health care system, and many hospitals had taken steps to reduce their ecological footprints by decreasing energy consumption and purchasing environmentally preferred supplies. However, there remained a critical need for understanding how patient care itself drives the demand for health care. By investigating the role of clinical decision making in the environmental sustainability of health care, we could begin to unravel the cultural forces that determine the ecological footprint of medicine.

Surrounded by stacks of foundational works on the sociology of death and dying,[2] along with cutting-edge research articles on the environmental aspects of health care facilities, I determined three primary reasons why end-of-life cancer care, in particular, offered a unique opportunity to investigate the role of patient care and medical decision making in the environmental outcomes of health care. First, end-of-life care is a high-cost medical setting: the 5 percent of Medicare recipients who die each year represent 30 percent of annual Medicare spending.[3] High-cost settings rely on high volumes of material and staff inputs, each of which has an ecological story to tell.

Second, over 1,600 Americans die of cancer every day, totaling nearly 600,000 cancer deaths each year. With such a relatively large number of people receiving end-of-life cancer care annually, the cumulative material and staff inputs required to provide cancer care provide insight into the environmental and public health consequences of the resources used to support cancer patients at end of life.[4]

Finally, there are three distinct end-of-life medical models that allow for comparative analysis: conventional curative care, palliative care, and hospice. Conventional care focuses on curing the patient of their illness and perceives death from disease as failure—a circumstance that helps explain the experiences I shared at the opening of this chapter.[5] Palliative care focuses on helping patients identify their quality of life goals when facing a life-limiting illness and, although not always the case, can help patients transition away from conventional curative care and toward hospice. Finally, hospice care supports quality of life for terminally ill patients by focusing on pain and symptom management. These three end-of-life cancer care settings lend themselves to comparative study, which is a useful approach for identifying when and where unintended environmental and public health consequences arise within systems. To begin understanding how the three settings differ in their approach to care, and how this leads to different volumes of medical supplies, pharmaceuticals, and other materials used to provide care, it is helpful to explore a bit of their history.

History of Dying in America

The hospital as an institution providing medical care to the dying is a relatively recent development.[6] In the late eighteenth and early nineteenth centuries, people dreaded going to hospitals for fear they would die from the poor sanitation, inadequate ventilation, and unhygienic practices of the untrained nurses who staffed the facilities. Throughout this period, most deaths in America took place at home in the company of family, friends, and neighbors, though the poor and indigent were sent to the hospital for care. It was not until the late nineteenth century that hospitals began transitioning into trusted centers of medical practice.

Several factors drove the transformation of hospitals. Scientific advancement, particularly the germ theory of disease, antiseptic surgery, anesthesia, and X-rays, which in part helped reduce mortality rates from nosocomial infections, dramatically shifted the public's perception of hospitals as places of healing. At the same time, industrialization moved a large portion of the population from rural areas to urban centers, where they no longer had the support of family to nurse them through illness. The professionalization of nursing, standardization of medical training, and transfer of hospital control to physicians and administrators rather than institutional trustees all contributed to making hospitals premier centers for medical care and training. These changes led people to entrust their health care to hospitals during acute periods of illness with the hope of receiving medical treatments that would return them to health.

In the mid-twentieth century, hospitals began providing care specifically for terminally ill patients. Again, industrialization played a role in this transformation since changing family structures, particularly the rural to urban transition, limited the ability for the dying to receive care at home. Meanwhile, scientific and technological advancement, particularly in the post–WWII era, offered increasing expertise and previously unknown chances to fight disease and prolong death, thereby serving to medicalize the dying process. Socially, death became a private, rather than public, affair during this period, and some historians argue that the hospital offered a way to shield the family from both the emotional burden they experienced at the death of a loved one and the grotesque nature of dying.

However, the displacement of the dying from home to hospital quickly raised a number of concerns. Medical staff began to see death as failure, and the aggressive use of medical technology to prolong life became commonplace. Sociological critiques reported that dying patients who were receiving hospital care experienced social isolation, anxiety, and mistrust of medical staff, often resulting from the widespread practice of withholding a terminal diagnosis from a patient. Thus opponents of the medicalization of dying began searching for other ways to support patients at the end of life.

Enter the modern hospice movement. Dame Cicely Saunders founded the modern hospice movement in London in 1967 with the opening of St. Christopher's in the Field hospice, an inpatient hospital specifically designed to care for the terminally ill, mainly cancer patients, by focusing on pain management and open communication among caregivers, patients, and their families. A few years later, Florence Wald, former dean of the Yale University School of Nursing, visited St. Christopher's and was so impressed by the model of care that she returned to Connecticut and opened the first U.S. hospice program in 1974. Bolstered by advocacy from hospice supporters, including Elisabeth Kubler-Ross, whose 1969 book *On Death and Dying* critiqued the highly medicalized approach to end-of-life care in the United States, hospice programs grew around the country. In 1986, following a four-year trial period, hospice gained the support of the federal government when Congress provided the hospice Medicare benefit to allow insurance to cover the cost of end-of-life care.

Despite the early success of hospice programs, hospitals continued to be the primary site of death for Americans suffering from chronic illness, and end-of-life care in hospital settings continued to draw criticism from both the public and health care providers. The SUPPORT study (the Study to Understand Prognoses and Preferences for Outcomes and Risks of Treatments) enrolled over 9,000 patients in five teaching hospitals across the United States to determine the appropriateness of care for terminally ill patients with an objective of reducing the number of patients who experienced painful, prolonged, and mechanically supported dying. The investigators found "substantial shortcomings" in the hospital care given to the terminally ill.

In response to the report, the Institute of Medicine (IOM) conducted a detailed study of end-of-life health care that culminated in the 1997 report, *Approaching Death*. The IOM report recommended several changes for improving end-of-life care, including the charge that, "palliative care should become, if not a medical specialty, at least a defined area of expertise, education, and research." As a result, palliative care units—with a goal of providing pain and symptom management to help alleviate patient suffering—began showing up within hospitals across the country. Currently, palliative care services continue to offer patients complex interdisciplinary care focused on pain control and symptom management, while also facilitating difficult conversations about life-limiting illness. Highly trained palliative care staff members support patients and families as they discuss their goals for care, including whether and when to opt out of life-prolonging measures.

With this brief history as our roadmap for understanding the different philosophies and goals that underlie the continuum of care available to patients as they near the end of their lives, we now turn to the three inpatient units that served as my research sites. In particular, the following descriptions offer a glimpse of the ways that the different philosophies of conventional care, palliative care, and hospice relate to clinical decision making, which in turn determines the cumulative impacts of care on the environment and public health.

Conventional Care: Learning to Roller Skate

After making my way through the long hospital corridors, stepping from the elevator, and pushing through the heavy set of double doors, I found myself in the conventional cancer inpatient unit at Hopewell Teaching Hospital. As I made my way to the main desk, nurses hurriedly traveled from room to room responding to patient calls, delivering medications, and changing IV lines. Medical students followed closely behind their attending physicians, while social workers and other support staff checked in on the progress of patients. Environmental services staff disinfected patient rooms and hauled red, black, and blue trash bags out of sight. The patients themselves were either resting in their private rooms or getting exercise in the hallways as they carried with them the tell-tale signs of cancer—shiny bald heads, bags of bright orange fluid dangling from glistening silver IV poles.

"I hope you brought your roller skates!" The nurses teased me as I headed out on the heels of Sarah, one of their compatriots. It was just before seven in the morning, and the day-shift nurses were arriving to begin their workday. The majority were waiting in the break room for the charge nurse to give them their patient assignments. Each registered nurse (RN) could expect to care for four patients, and each nursing assistant was assigned to work with six patients. A total of thirty beds on the conventional cancer unit meant that seven RNs and five nursing assistants worked on days when the census was high and there was a patient in each room. On this day the unit was full, and busy.

Sarah zipped through the hallways, checking in with night-shift nurses for updates and insights on her assigned patients, then logged into the electronic medical record system to make brief notes to help her remember timing for her patients' medications and scheduled procedures. It was the only time she would sit all day. As I followed closely behind Sarah, asking questions when time permitted and jotting down observations of the materials used to provide patient care, I really did begin to wish I had roller skates.

The conventional unit itself held thirty small, private patient rooms, each with its own bathroom and the standard disinfectant "hospital smell." The sounds of televisions clicking through channels, feet hurrying down hallways, and machines registering alarm filled the air. Here among the hustle and bustle, the ultimate goal of care was to cure patients of cancer. The majority of cancer patients at Hopewell were treated on an outpatient basis, which provided the double benefit of lowering hospital costs while also keeping patients at home where they preferred to stay.[7] But when a patient's symptoms did not respond to the treatments offered through the outpatient clinic, the clinical staff would refer the patient upstairs for admission into the inpatient unit. As a result, the unit provided care to patients through three different cancer services: hematology, bone marrow transplant (BMT), and oncology.

The hematology service treated patients with leukemia and lymphoma—cancers of the blood. These patients were typically admitted to the unit to receive inpatient chemotherapy treatments, and a number participated in clinical trials of new chemotherapy drugs that kept them on the unit for a few weeks. Patients undergoing BMT were also admitted to receive treatment for blood cancers, but typically their diseases had not responded to the standard treatments offered through the hematology service, so they were on the unit for a more aggressive treatment regimen. These patients first received high-dose chemotherapy to wipe out all their bone marrow and then received an infusion, or "transplant," of new, nondiseased bone marrow. The process was arduous, and BMT patients could expect to be on the unit longer than all other patients—up to a month. In contrast, patients on the oncology service who were suffering from solid tumors of the breast, lung, colon, pancreas, brain, and other organ sites were usually admitted for only a few days. Generally, these solid tumor cancers were treated with chemotherapy at the outpatient clinic, but patients whose symptoms worsened would be admitted to the inpatient unit for management of pain and other symptoms of their disease.

Many of the patients on Hopewell's conventional cancer unit had been admitted as a last chance for a cure. One administrator at Hopewell described the patient population: "It's rare for people to come into the hospital for cancer treatment. We try to keep people out of the hospital—it's not desirable to be in the hospital from a social standpoint, economic, even a health standpoint. People who are admitted, that's their only alternative. If care could be provided in any other way, it would be."

Of the fifty-three cancer patients whose care I observed over the year and a half I spent on the conventional unit, sixteen had a hematological diagnosis, fifteen were admitted for BMT, and twenty-two were oncology patients. The majority of care I observed on the hematology and BMT services was for treating patients with acute myeloid leukemia, which had a 28 percent five-year relative survival rate, meaning that just over one quarter of people diagnosed with acute myeloid leukemia are expected to be alive five years after their diagnosis.[8] Other hematology patients whose care I observed had diagnoses with more favorable prognoses: the five-year relative survival for non-Hodgkin's lymphoma is 72 percent, and 87 percent for Hodgkin's disease. Diagnoses and prognostic expectations for oncology patients are highly variable and depend on the disease type and the stage at which the cancer is first diagnosed. On average, the five-year relative survival rate for breast cancer is about 90 percent, 64 percent for colon cancer, 33 percent for brain cancer, 19 percent for lung, and 9 percent for pancreatic cancer—all diagnoses I observed at Hopewell. However, some of the patients were admitted to the inpatient oncology service with a variety of complications, which rendered their individual prognoses less favorable than these figures suggest.

"I heard some statistic that from the time our oncology patients were first admitted to the unit until time of death was less than six months." Joanne, one of the nursing administrators on the unit, sat across the desk from me, scrolling through email messages on her computer screen as she spoke. "When you're admitted to the unit as an onc patient you're really at the end-stage of your disease. You're there for symptom management and complications. If you're an onc patient and you're doing well you're in the [outpatient] clinic. . . . So I think that characterizes our onc patients as having a lot of symptom management and a lot of decline by the time we see them."

According to the American Society of Oncology Clinicians, the use of aggressive, curative treatments, such as surgery, chemotherapy, and radiation, among patients with incurable cancers causes more harm than benefit to patients.[9] Despite this fact, national-level data show that patients continue to receive aggressive curative treatments right up to the point of death. The impetus for the continued use of such "futile" treatments varies, and we will explore these social drivers further in the coming chapters. For now, the key point is that many patients who are within the final six months of life receive aggressive, curative cancer care—and those treatments come at a cost to the patient, their caregivers, society, and the planet.

Palliative Care: The Quiet Hallway

The social and financial costs of aggressive treatment among people who received more harm than good from the conventional approach helped lead to the development of palliative care services within many hospitals, including Hopewell. During my first day observing staff on Hopewell's palliative care unit, I was surprised by how quiet the unit felt. Just thirty feet away, the nurses' station of the conventional cancer unit was a hub of activity with clinical staff, physicians, administrators, patients, visitors, and housekeepers circulating through to speak with patients or transport them to procedures, draw blood samples, or deliver supplies. After that short walk down the hallway, I pushed through the double doors that separate the commotion of the cancer unit from the palliative care unit and found myself surrounded by stillness. This sense of quiet reflected the difference between the goals of the palliative care unit and the conventional unit. Hopewell Hospital's ten-bed inpatient palliative care unit was developed in direct response to the internal study that found up to 75 percent of patients died within six months of being admitted for inpatient oncology care. The new unit was designed to be a place of quiet and calm where patients and family could be spared the regular commotion of the hospital.

Echoing national sentiments, the development of the new unit made sense both socially and economically.[10] Socially, palliative care benefits patients by offering a focus on quality of life through goal setting—for example, whether the patient would prefer to be at home—and pain and symptom management. Financially, palliative care offers a lower-cost alternative to allopathic care.[11] The lower cost reflects a decreased reliance on tests and procedures, which in turn means a lower demand for costly clinician time and medical supplies.

Hopewell's palliative care service included physicians, social workers, nurse practitioners, and case managers who provided care through consultation with specialty teams throughout the hospital, such as oncology, intensive care, and cardiology. During my time at Hopewell, I observed the care of twelve cancer patients on the palliative care unit.

Dr. Smith described an initial palliative care consult as an opportunity to assess the goals of a patient and their family. When he first sat down with them, he would ask, "If things take a turn for the worse and it becomes more difficult to control the symptoms, such as pain, would you [the patient] prefer to be more comfortable by taking some medication knowing that it might make you sleepy and harder for you to communicate with your loved ones? Or would you prefer to be alert and talking with family and friends even if that meant having a bit more pain?" This information is helpful to a patient, whether they are trying to decide about continuing life-prolonging treatment or not. He explained, "Depending on the circumstances, sometimes the questions instead turn to asking whether the person would prefer focusing on a certain quality of life or rather on lengthening life as much as possible."

Through these initial palliative care consultations, the team helped patients and families articulate their goals—to go home, to be closer to home in a skilled nursing facility—and would begin to craft a plan for managing symptoms to help accomplish those goals. In many cases, the patient would remain on the original unit where they had first been admitted, and the palliative care staff would visit to oversee care until the new goals were met. This was one reason why the palliative care unit itself was quiet—many beds were often empty. In other cases, patients were transferred to the palliative care unit when the clinicians were all in agreement that the move would better meet the patient's and family's needs—and in certain cases, the needs of the hospital to open up beds in other units, particularly the intensive care unit.

Carolyn was a nurse who split her time between the conventional and palliative care units. She described the work of the palliative care team this way: "They really try to get at what that person's and family's goal is—whether it's to go home with hospice or, some patients want to stay in the hospital, that's where they feel safe and that's where they feel cared for. That's not always possible because of insurance. And so sometimes you run into problems where they want to stay but they really might not die for another few weeks and we really can't justify keeping them in the hospital just to die. Which sounds kind of cruel, but I know the palliative care service has a length-of-stay where their focus is really to get people in, manage their symptoms, and get them out of the hospital—whether that was home with hospice, or a nursing home with hospice, wherever was most appropriate. But it wasn't to replace hospice. It was to tune people up, get their symptoms under control, and get them where they needed to go." For those patients who were discharged to hospice and were in need of acute medical care, many were transferred to Baluster Hospice just a few miles down the road.

Hospice: The Cadillac of Care

"In hospice, we say there's no such thing as an emergency." Cindy, an RN team leader walked me through the hall on the way to one of the nursing stations to begin the day shift. "Even though our patients are very ill and need very intense care, it's different from the care you would give in a hospital."

On our way from the front entrance to the acute care inpatient unit at Baluster Hospice, Cindy's sentiment rang true as I noticed a quiet sense of busyness. Staff members passed through the sun-filled hallways individually or in pairs, glancing out the windows to check the weather. Newly arriving patients were wheeled to their assigned rooms as staff members prepared paperwork for their admission. Visitors checked in at the front desk, then made their way to patient rooms. While I had found the palliative care hallway almost uncomfortably silent, the hospice facility was filled with calm activity.

Staff here referred to the facility as the "Cadillac of care." The reason for the nickname became apparent when I began following staff into patient rooms. Each of the twenty-four private patient rooms at Baluster offered a view overlooking a private patio complete with outdoor seating area, bird feeder, and a view of the woods beyond. French doors—wide enough for a hospital bed to pass through—connected the indoor space to the sunshine and breeze outdoors. Gazing around the spacious rooms themselves, the hardwood floors and brightly colored walls adorned with artwork made it easy to imagine this as a country guesthouse rather than a medical facility. A closer look revealed the medical features of the rooms—electronic hospital beds, oxygen pumps hiding behind the paintings, handrails in the bathrooms—but for the most part, the facility was designed to feel more like "home" than "hospital."

The immediate differences between Hopewell and Baluster stemmed from their different goals of care. Cindy described hospice by saying, "The goal of care here is very patient and family driven. We focus very closely on what the patient wants—or the family wants if the patient is unable to speak for themselves. It's very comfort-centered, as opposed to curative."

Dr. White, one of the hospice physicians, expanded on the idea of comfort by saying, "Well, the goal here is the level of comfort the patient wants to choose. Some people do choose to be in pain." Particularly, patients who want to continue communicating with their family will often choose having some degree of pain if it means they will be alert enough to converse.

Dr. White added, "I think of that as quality of life. You don't want to impair their mind at this last end of their life, and that frequently happens with medications."

Overall, the goals of hospice to provide patient- and family-centered comfort care sound similar to the goals of palliative care, but there are a few important differences. For a patient to qualify for admission into a hospice program, a physician must certify that the patient has a maximum prognosis of six months if their disease runs its normal course, and while they can receive medical treatment for pain and symptom management, they must forgo curative treatments.[12] This latter difference helps explain, in part, the creation of Baluster Hospice's acute care inpatient facility. According to one of Baluster Hospice's founding members, before the unit existed Baluster functioned as a home-based program. At that time, any patient whose acute pain and symptoms could not be managed at home was admitted to Hopewell Teaching Hospital until their needs were met and they could be discharged back home. With this arrangement, however, it was possible that the patient would undergo "curative" diagnostics and tests in the hospital that did not fit with the hospice philosophy, nor, perhaps, with the patient's wishes. This philosophical dilemma, combined with the financial arrangement where Baluster Hospice paid Hopewell Teaching Hospital a per diem rate for each admitted patient, led to the establishment of Baluster's inpatient facility where costs could be better controlled and patients could be spared unnecessary treatment—with the hopeful goal of offering these patients a better quality of life.

Of the fifty-six patients whose care I observed at Baluster Hospice's inpatient facility, twenty-six had a primary diagnosis of cancer, and the remainder suffered from other illness, including end-stage renal disease, chronic obstructive pulmonary disease, chronic heart failure, and dementia. And while hospice is of course associated with dying, only twenty of the fifty-six patients whose care I observed were admitted for end-of-life care. The remaining patients were there for symptom management with a goal of returning to their place of residence after discharge.

Cumulative Costs of Care

The different philosophies and goals of care being practiced in these three settings required different material resource inputs, which in turn led to different environmental and human health outcomes. As one oncologist at Hopewell stated, "Conventional care means more imaging, more blood tests, chemotherapy oftentimes. And anytime you have an intervention that you do, it means everything else that's associated with it—blood [samples], and scans, and X-rays, and catheters, and IVs, and you know, everything else. And there are just more interventions with conventional care; so more interventions means more of everything else." This statement proved true in my observations and reflected the overall trend of ecological impacts arising within these three sites: more interventions in conventional care did mean more of everything else, including a larger quantity of supplies used, amount of pharmaceuticals administered, and volume of waste generated. While each of these three medical sites had ecological impacts that should be taken into account, both my observations and the purchasing and disposal data available for verifying my observations show that overall, the medical treatment provided on the conventional cancer inpatient unit had greater cumulative impacts than either the palliative care unit or the hospice unit.

The following chapters describe these impacts in detail by telling the stories of the people and materials I encountered on my journey through these three medical inpatient units. These stories provide a glimpse of providing medical care day to day, follow the global flow of materials and natural resources used in patient care, and offer insight into how to overcome these impacts while hopefully also providing options to people that support their preferences for dying. One key insight I have kept in mind is that while 60 percent of terminally ill cancer patients report that they would prefer to die at home, only 42 percent actually achieve this goal.[13] As a result, the quest for medical care that is better for the planet may begin with identifying care that is better for people.

Length of Stay as a Key Driver of Cumulative Impacts

Many of the hospice staff members I worked with enjoyed telling stories of patients who had received such benefit from hospice services that they had well outlived the required six-month prognosis that made them eligible for hospice services. Some people had stories of patients who had improved so much they were no longer eligible for hospice. Despite these anecdotes, the majority of patients who transfer to hospice programs across the nation die within just one month of transitioning to hospice care.[14] When I asked why patients have such a short length of time with hospice despite the fact that they are eligible for a much longer period of time, I always heard the same response. "Sometimes it's the doctor, sometimes it's the patient, and sometimes it's the family" who simply are not ready to give up the hope that conventional life-prolonging treatments offer. And to sign on to hospice means that the patient has forgone further curative treatments. But everyone I asked—whether on the conventional, palliative care, or hospice team—suggested that being on hospice for a longer period of time can provide a number of benefits to both the patient and their family. This limited length of stay of terminally ill patients in hospice may reduce these benefits and brings to question how the environmental impacts may differ between patients who remain in intensive material resource use curative care compared with palliative care and hospice settings that may have lower reliance on material inputs.

According to Samantha, one of the nurses I worked with at hospice, signing onto hospice earlier means that patients are "going to have a better quality of life, they'll have support from all of the different team members, the interdisciplinary team members, they'll have support in their home—social workers, grief counselors, nursing assistants, nurses, chaplains, all the disciplines."

A nursing administrator for Baluster who had worked at conventional care facilities in the past said, "We don't do as many labs here. I'm not worried about the very minute details to try to tweak something to try to get someone better. You're really looking at more of a global picture of the patient." Taking a more global view was exactly what I had come here seeking, though from my perspective, that global view would soon have me following medical supplies, pharmaceuticals, and waste as they flowed through each of these three medicalized settings.

2

Medical Waste

Early one Friday morning, while the hallway lights were dimmed and the muffled voices of staff members were the only noises heard, a patient on Hopewell Hospital's palliative care unit died. A few hours later, after the family had left and the nursing staff had removed the patient's body, Alex, the housekeeper, arrived to do a "discharge cleaning" of the room. Within the next hour, Alex would clean and disinfect the entire room from floor to ceiling in anticipation of the next patient's arrival.

The room was full of stuff. The patient's teddy bear lay haphazardly on the windowsill along with a number of medical devices the family had brought from home. Alex placed these personal belongings in a bag to return to the patient's family. Next came the medical supplies strewn about the room. There were boxes of unopened gauze and bandages, a handful of unused medical gloves that had fallen out of their cardboard box, and two unopened, prefabricated medical kits for various procedures. As he moved around the room, Alex filled a 30-gallon black plastic garbage bag with unused medical equipment—all the items that had been brought to the room in the flurry to help treat and comfort the patient in her final hours of life—and it would all be landfilled. Hospital policy considered any items that entered a patient's room to be potentially contaminated, so the bag full of unused supplies was promptly trashed.

Over the next few minutes, Alex filled two additional 30-gallon garbage bags with used materials, including IV bags, plastic packaging, medical gloves, pink plastic basins and water pitchers. After packing up this garbage, Alex moved swiftly around the corners of the hospital bed, tugging off the sheets. He quickly filled a clear plastic bag with these linens in addition to soiled towels and gowns, then opened the linen closet that was brimming full of freshly laundered materials and pulled down the neat stacks to add to a second clear plastic bag. These items would be relaundered, even though never used, since the hospital's infection control standard required such precautions, again to avoid spreading pathogens to the next patient.

Alex, like his fellow housekeepers, wore a pair of synthetic blue nitrile gloves along with his uniform—a light blue button-down shirt and navy pants—to protect himself from any potential pathogens he would encounter in his work. The soiled linens brushed against Alex's bare arms and clothes as he bundled them together, but he paid no concern to whatever invisible contaminants might be in the sheets. He was in a hurry. He needed to get this room turned around for the next patient as quickly as possible.

After removing the laundry bags to the hallway and perching the bags of garbage against the edge of his cart, Alex was ready to call his supervisor to officially report that he was now cleaning the room. Once he called it in, he was allotted only forty-five minutes to disinfect the entire room. He had decided to bag up the patient's personal belongings, the garbage, and the linens before calling in the official start time because he worried that there was so much stuff he would not be done in the required time. He dialed in the number at 10:44 a.m. "Go time," he told me. With precision and efficiency, Alex cleaned and disinfected the room from top to bottom and wall to wall. In effect the entire room—including the bed, chairs, and bathroom—was dusted, mopped, and washed. At 11:27 a.m., forty-three minutes after officially starting his task, Alex called in to report that the room was clean and ready. After one last glance around the room, he flicked off the light switch and rolled his cart down the hallway to the soiled utility room, where he placed all the used materials. He piled the laundry bags into one cart and the garbage into another.

As he finished throwing the last bag of trash, I asked where all these materials would go next, just as I had asked many other staff members. And just as the others had replied, he told me it went to the basement, but beyond that he did not know. As he swung the door closed behind us, we agreed it would be interesting to find out.

The Five Waste Streams of Health Care

The majority of people I shadowed at both Hopewell Hospital and Baluster Hospice commented about the vast amount of waste generated in their work, and yet no one who worked on the patient care floors knew the next steps involved in the fate of the waste. Housekeepers neither knew where the various waste streams went nor what happened next in the disposal process. The nursing staff and physicians—who are less intimately involved than the housekeepers in disposing of the numerous waste streams—certainly did not know. Yet everyone I asked about the fate of these materials expressed interest in having that knowledge. My goals for sharing my insights on waste are therefore to shed light on the actual fate and unintended consequences of the various waste streams on the environment, workers, and public health, and to examine the social and economic factors that govern the movement of waste through these medical settings. In addition, this chapter offers insights into best practices for minimizing the negative consequences of hospital waste streams—including some that are arguably simple, and others that require more complex changes.

There were five waste streams at both the Hopewell Hospital and the Baluster Hospice facilities, including confidential paper, pharmaceutical waste, municipal trash, infectious waste, and single-stream recycling. The first waste stream, confidential paper, followed the same process at both Hopewell and Baluster in order to comply with the Health Insurance Portability and Accountability Act, which safeguards patients' personal information. Confidential paper was stored in locked containers, then picked up, shredded, and recycled by special confidential waste stream haulers. Paper recycling offers many environmental and economic benefits over both incineration and landfilling, including a reduction in landfill space needed for disposal and reduced greenhouse gas emissions from methane produced in landfills as the paper fiber decomposes.[1] Furthermore, every ton of paper that is recycled saves 7,000 gallons of water and an equivalent amount of energy to power a home for six months when compared to harvesting trees for virgin paper. The public health outcomes of paper recycling are less clear, with some concern over the exposure of workers to caustic chemicals in the de-inking process,[2] but overall the paper recycling process employed at both Hopewell and Baluster appeared to have minimal environmental and public health impacts.

In contrast, I discovered such an array of unintended consequences associated with pharmaceutical waste that I devote an entire chapter to these therapeutic materials and their disposal. The final three waste streams—municipal trash, infectious waste, and single-stream recycling—I report on in detail in this chapter. In addition to being of interest to the staff members on the inpatient units, the sheer volume of waste produced within the health care sector suggests that examining the factors that govern the creation and disposal of waste is necessary for identifying points of intervention in order to increase the environmental sustainability of medical care.

Over the course of one year during my observations, Hopewell Hospital disposed of a combined total of nearly 2,600 tons (5.2 million pounds) of municipal trash, recycling, and infectious waste, while Baluster Hospice disposed of almost 88 tons (176,000 pounds) of these waste streams. Since Hopewell is a much larger facility, serves more patients, and requires a larger staff, we might expect it to produce more waste. However, when we apply a different metric to describe the volume of waste generated in each facility a more nuanced understanding of waste emerges. For this purpose, I calculated the average amount of waste created per patient day in order to compare waste generation at these two facilities.

A "patient day" represents the total number of days that the total number of patients were in a medical facility. For example, 100 patients in a hospital for one day equals 100 patient days. The average amount of waste produced at Hopewell Hospital was 30 pounds per patient day, compared to 14 pounds at Baluster Hospice. In contrast, each person in the United States produces about 4.5 pounds of trash and recycling combined each day.[3] Compared to the general public, these medical facilities created between four and a half to more than nine and a half times the amount of waste that people produce in their homes. Importantly, these medical facilities also differed in the volume of each type of waste produced. With regard to municipal trash, Hopewell produced an average of 20 pounds per patient day (67 percent of total waste) compared to 10 pounds (71 percent) at Baluster. For single-stream recycling, Hopewell produced an average of 6 pounds per patient day (20 percent) compared to 4 pounds (28 percent) at the hospice. Infectious waste generation included 4 pounds (13 percent) at Hopewell, compared to less than one-tenth of a pound (less than 1 percent) at Baluster.

Since there were no data available on the amount of waste produced on my specific inpatient units of interest, please note that these numbers represent the waste coming out of each entire facility—including the cafeterias, administrative offices, and other areas.[4] At Hopewell Hospital this included the emergency room, surgery, diagnostic laboratories, and other units that had no equivalent at Baluster Hospice, so it makes sense that the hospital produced more infectious waste than the hospice facility. At the same time, we should not dismiss the differences in waste generation of these facilities on the basis that they do not each offer the same services since these differences reflect their individual philosophies of medical intervention.

Municipal Waste

The comparative impacts of Hopewell and Baluster's municipal waste, single-stream recycling, and infectious waste streams varied depending on the next steps in their disposal pathways. In the case of municipal waste—basically household trash—the primary environmental and public health concerns arise from air pollutants and greenhouse gas emissions from both transport of these materials to the landfill and decomposition within the landfill. Concerning waste transport, diesel trucks remain the primary option for hauling waste from its source at the health care facility to the disposal site. Along the way, these trucks emit a number of air pollutants regulated by the U.S. Environmental Protection Agency, including nitrogen oxides, sulfur dioxide, carbon monoxide, and particulate matter. These pollutants affect public health both directly, such as in the case of exposures to particulate matter, and indirectly through the health-related impacts of climate change.[5] Importantly, communities living along diesel truck routes tend to be communities of color and lower income, which brings concerns over the impacts of heavy-duty truck transportation into the realm of environmental and social justice.

In practical terms, the keys to reducing the impacts of health care waste heading to the landfill are to decrease the volume of waste produced and the distance the waste is hauled. Every pound of trash disposed of via landfill emits 0.94 pounds of methane and carbon dioxide—both potent greenhouse gases—as the waste decomposes.[6] Of note, both landfills where Hopewell and Baluster sent their waste captured methane from the decomposition process and sold it to their local communities for electricity generation, thereby helping to curb the harms posed by emissions. However, all landfills have limited life spans, thereby presenting an intergenerational dilemma as the environmental burden of waste disposal shifts to future generations who will face the eventual release of pollutants and leachate into the surrounding environment over time.[7]

Another challenge posed by landfills is that they eventually reach capacity and are closed, requiring the siting of new facilities. Very few communities are willing to have a landfill located in their neighborhood, a community response typically referred to as "not in my backyard," or NIMBY. As a result, trash may be hauled increasingly longer distances (at increasingly higher costs) for disposal, further adding to problems of air pollution and carbon emissions. The greater the total distance traveled, the greater the unintended environmental and public health consequences. Baluster Hospice contracted with a local hauling company to transport its municipal trash back to the company's transfer station facility. Here the waste was compacted, loaded onto semitrailers, and hauled to the company's private landfill in an adjacent county. In total, the 63 tons of municipal waste generated at Baluster each year traveled about 86 miles from the health care facility to its final disposal in the landfill. That equals over 5,400 cumulative miles (63 tons × 86 miles) traveled.

In contrast, a different local hauling company handled the municipal trash at Hopewell Hospital and transported this waste to the public municipal landfill about 13 miles away. Hopewell Hospital generated over 1,900 tons of municipal waste, totaling 24,700 cumulative miles of hauling—about 4.5 times greater than the cumulative trash transport of Baluster. This suggests that the emissions from municipal waste transport were 4.5 times greater for the hospital than for the hospice facility.

The research describing the public health impacts that result from municipal landfills is somewhat limited. The clearest evidence is that children living in close proximity to landfills have elevated lead levels as compared to the general population.[8] In addition, several studies have suggested that people who live within the vicinity of landfills have higher rates of congenital birth defects, low birth weight, and cancer incidence, but to date these studies are typically flawed by confounding factors, and further research is needed to tease apart the complexity of measuring health outcomes on communities within the vicinity of landfills. For example, since landfills are typically undesirable places to live near, they are often cited in low-income, minority communities whose health problems may reflect poor access to health care, nutrition, and other factors rather than potential landfill exposures. However, this point again raises the environmental justice concern that undesirable facilities, such as landfills, are disproportionately sited in communities of color and low-income communities.

Social Justice and the Environmental Services Workforce

I encountered a similar question of social justice as I waded further down the waste stream. At Hopewell and Baluster, members of the Environmental Services staff who handle the various waste streams were much more likely to be people from marginalized communities or new immigrants than any other staff group I worked with—a fact that reflects larger national trends.[9] At Hopewell, over three hundred people made up the Environmental Services Department. About 47 percent of these workers were Caucasian, 23 percent were Latino, and 21 percent were Black (both African American and new immigrants). The majority of the remaining 10 percent of staff members were recent immigrants who were ethnic minorities in their home Asian countries. In comparison, of all the nursing staff I was in contact with on the conventional and palliative care units, only about 5 percent belonged to underprivileged communities. At Baluster Hospice, the waste-handling staff was much smaller since the facility itself was much smaller, but there too the housekeeping staff comprised a different population than the nursing staff. In my observations, all the housekeepers at Baluster's acute care unit were new immigrants or from marginalized communities, compared to only a fraction of the nursing staff.

For about 35 percent of Hopewell's housekeeping and decontamination staff, English was a second language. In my observations of a training session at Hopewell designed to test staff members' knowledge of proper recycling practices, many had trouble reading the material, and, according to one of the trainers, many were dyslexic or had other cognitive difficulties. This is concerning because if staff members did not fully understand the training materials they were given, they could be at higher risk for exposure to hazards in the workplace.

"These aren't your cream of the crop students," the trainer told me. "These are the kids who got Bs and Cs and Ds in school, if they finished school at all." This work paid comparatively well for someone without a strong educational background—starting pay was $11.44 per hour when I interviewed some Environmental Services administrators and was about to increase. This compared well to the U.S. mandated minimum wage of $7.25 per hour.[10] Thinking back to the administrator on the conventional care unit who told me that cancer patients were admitted to the hospital only when they had no other option, it could be easy to draw a comparison here—working with medical waste streams is a pretty good option for people who need a job, even if it means possible health exposures.

Invisibility and the Imperfect Separation of Waste Streams

As I followed Mary, one of the housekeepers on Hopewell's conventional cancer unit, I asked her if she knew where all the garbage went, just as I had asked Alex a few days earlier. We were on our way to deposit a few bags of municipal trash in the soiled utility room, and Mary replied, "I'm really not sure. You should follow someone from decontam." After a moment of slinging the garbage bags into a pile, Mary paused, and her careworn face suddenly became animated, "I bet it would be like Alice going down the rabbit hole!"

We were standing amid all the morning's waste from the conventional and palliative care units. Several rolling carts and bins lined the walls of the room, ready to receive the used materials. The various bins and containers were quickly filling and were nearly ready to be whisked away, but to where, exactly, no one seemed to know. Industrial-looking gray elevator doors stood in the back corner of the room, silent at the moment.

After cleaning a few more patient rooms, Mary took her fifteen-minute break. We sat in a small hallway area just outside the soiled utility room, and as I asked her a few more questions about her work, one of Mary's coworkers walked by and waved hello to us.

"He might know where all the stuff goes, he's from decontam." Despite the fact that Mary did not know her coworker personally, she recognized him from the daily morning meeting attended by all the Environmental Services Department employees, and she knew his job was to transfer materials from the soiled utility room to the basement. Mary waved the man over and told him about my project, asking if he knew where all the garbage went in the basement. He nodded excitedly, and since I was supposed to meet up with another Environmental Services employee in the basement in a few minutes anyway, in broken English he kindly agreed to escort me to "decontam." As we made our way back into the soiled utility room, he humbly apologized for his emerging English language skills. In return, I humbly apologized for my limitation of speaking only English. The next minute we were waiting for the previously silent utility elevator doors to open and carry us to what I would learn the decontam staff affectionately called the "armpit of the hospital."

Once I reached decontam, I met up with Alice, who had agreed to show me the ropes down in the basement. A moment later, I was following Alice's fast-moving heels through a maze of hallways and rooms on our way to the loading dock where all the municipal trash and recycling was eventually gathered. We passed through rooms where workers donned white hazmat suits as they cleaned and disinfected materials for reuse throughout the hospital. The air felt heavy, wet, and warm from the steam cleaners and power washers in the rooms, and it had an overpowering acrid smell. I clutched my clipboard and tried to avoid coming in contact with anything we passed. When we finally made it outside to the loading dock, I found myself standing near a giant garbage compactor as Alice bent down over a small pile of black plastic garbage bags; the color meant that these bags were part of the waste stream destined for the municipal landfill. She opened a few bags just to show me what was inside, just as she would if she were training me as a new hire. She told me that her point in this exercise was to make people aware that what they thought they were handling might be something very different. She suggested that the staff here should beware; a presumed water droplet that dripped from a bag and landed on a worker's arm may in fact be something much less innocuous than water.

Inside the first bag, we found a plastic drape from a procedure, a heart catheter, gloves, and empty plastic syringes. Alice said that this one bag full of stuff had likely come from a single procedure on the cardiac transplant unit. The second bag contained gloves, suction tubing, blue paper drapes, gauze, a green plastic handle used to turn a machine on without producing a spark, and dripping saline IV bags. The third bag simply held suction tubing, but the fourth bag won the prize; inside was suction tubing and a clear plastic suction container topped with a bright green lid—and it was full of bloody phlegm and fluid. Wearing her gloves, Alice hauled the fluid-filled container out of the bag, tied the bag back up, and tossed it into the compactor along with the other noninfectious municipal garbage. Then we headed to the "swamp" to dispose of the infectious waste properly—in the *red* bags. Alice seemed unfazed by what we had found. She said that it happened all the time, especially when the task of disposal fell upon new staff members who had not been properly trained.

Peering into the bags on the loading dock, I gleaned some important insights into how to reduce the negative impacts of medical waste streams.

Here, the color of the garbage bag signaled its contents, where it should go next, and how it should be processed. Black bags signified municipal trash, thereby providing information to the waste handlers about the contents and relative safety from health exposures. Red bags signified infectious waste and therefore higher risks of exposure. This information was important to housekeepers, decontam staff, waste haulers, and landfill workers down the line because it immediately conveyed relative risk.

The color-coded red bag system resulted from federal mandates in the Medical Waste Tracking Act of 1988, which classified infectious waste as a regulated entity after improperly disposed of medical waste began washing up on the shores of several East Coast states.[11] Red bags displaying the biohazard symbol became an easily recognizable way to signal the difference between municipal trash and infectious waste, and they helped protect hospitals from substantial fines levied when infectious waste ended up in the wrong place. Infectious waste is not simply disposed of along with municipal trash because it carries a much higher public health risk. Although infectious waste is more expensive to dispose of than municipal trash, the improper disposal of red bags carries fines that incentivize proper separation and handling. A number of people at Hopewell alerted me to the fact that years earlier a landfill employee had been stuck by an improperly disposed of needle, and as a result, the hospital was responsible for the employee's annual HIV testing and any needed treatment.

The color-coded system was designed to reduce the risk of exposure to infectious agents among waste-handling staff. However, as my observations with Alice point out, black bags also created a barrier to knowing what was inside. Since the contents of the bags could not be seen, it was impossible to know whether they actually held the noninfectious waste they were meant to contain. Since Alice and her coworkers only opened bags to train new hires about the possible risks of handling waste, the majority of bags traveled unopened to their final disposal site, potentially exposing a number of people along the way to any infectious waste they may have held.

How can we minimize the risks of exposure to improperly disposed infectious waste in the municipal waste stream? There are a number of governing factors that might explain why the infectious container that Alice discovered ended up in the wrong waste stream. Maybe the nurse who disposed of it was new, as Alice suggested, and had not been properly trained yet. Or perhaps the nurse was in a hurry because her other patients needed immediate care, but she had to clean this room before she could respond to their calls. In any case, the infectious waste was diverted from the proper waste stream, thereby introducing a number of possible exposures to staff members and waste haulers down the line.

Proper waste disposal training should occur prior to a new staff member working on a unit to prevent the flow of infectious waste into the municipal waste stream. But if a nurse is from a temp agency, proper waste-handling instructions may be overshadowed by learning the computer system. In the case of a lack of time, perhaps having a lower nurse to patient ratio on the unit would allow the RNs the time needed to properly dispose of materials while also meeting their patients' needs. In my observations on the conventional and palliative care units, the RNs were responsible for cleaning up bodily fluids. Both nursing assistants and housekeepers would track down a patient's primary nurse to let them know when there was such a cleanup task that required their attention. Given that the RNs were typically the busiest staff members on the units—not having time to sit down, eat lunch, and in some cases take a bathroom break—perhaps other staff members could receive training in the proper disposal of bodily fluids and infectious waste. Such a reallocation of responsibilities could help reduce the RN's time crunch and increase the probability that waste is disposed of correctly.

However, social disconnections between various staff positions may also be responsible for the breakdown in the system. To run a hospital the size of Hopewell efficiently, separation of roles and responsibilities is necessary. The utility elevator in the soiled utility room is the conduit for decontam workers, but not for the housekeeping staff who pile the various garbage bags in bins, nor for the nursing staff who are responsible for cleaning up after procedures. The sheer scale of operations at Hopewell required people to function within their specific designated areas—which prevented staff members from various departments from forming relationships with one another. This social disconnection creates invisibility within the waste-handling process and builds a barrier between the clinical staff who work at the point of waste generation and the Environmental Services staff who are downstream.

Having knowledge of the person who will be handling—and exposed to—waste may increase a relational obligation between staff members and decrease the amount of improperly disposed of materials. This possible explanation dawned on me when I realized that even on the conventional and palliative care units where the same nurses and housekeepers work side by side every day, few if any of these staff members knew one another's names and hardly ever acknowledged one another. The lack of social connection between staff members at Hopewell became apparent to me only after I had spent some time at Baluster Hospice where, in a much smaller facility, the nursing staff and housekeepers greeted one another by name each morning. The complexity of mitigating any of these possible explanations—lack of staff training, lack of time, lack of social connections—of course depends on the political economy and institutional values of the medical facility.

In my observations at the hospital's conventional cancer and palliative care inpatient units, I never witnessed any infectious waste entering the municipal trash stream. The majority of black bag trash on the units consisted of gloves, pill packaging, empty saline bags, IV tubing, and the like. So the chances that the municipal waste from these units would expose workers downstream to potentially infectious substances seemed relatively small. But once the housekeepers removed the trash from patient rooms and placed it in the soiled utility room, it was mixed with the black bags from units and departments throughout the hospital.

Alice and other decontam staff amassed trash from unit to unit and floor to floor as they traveled up and down the utility elevators collecting the various carts and bins to take to the basement. By the time the trash from the cancer and palliative care units reached the dumpster at the loading dock, it had been mixed with black bags from throughout the hospital. While places like the cafeteria and administrative offices were unlikely to add infectious waste to the municipal trash stream, black bags containing bodily fluids from the cardiac unit and other patient care areas could potentially contaminate the trash, thereby exposing workers downstream.

According to administrators at Hopewell and at the waste hauling company, there had been times when landfill workers had reported finding red bags mixed in with the municipal trash. When a violation like this occurred, there was a system of governance in place to prevent the problem from recurring. Along with possible fines, the landfill required a hospital employee to accompany the trash to the landfill, "basically until the landfill operator is satisfied the issue has been addressed internally by the hospital—no red bags over a thirty-day time period is typical." These measures both help protect the health of landfill staff and effectively ensure that the hospital monitors its disposal practices to correct any problems in the system. However, no one commented on the potential for infectious waste to make its way to the landfill in black bags. People only spoke of the ramifications for red bags entering the municipal trash stream, rather than consequences of infectious waste riding undetected in the black bags.

At Baluster Hospice, I observed a much lower probability of infectious waste contaminating the municipal trash simply because the activities in the facility did not produce much red bag waste. The majority of infectious waste at the hospice facility was in the form of "sharps," which had special disposal containers to minimize needle pricks. Since Baluster had no operating room, diagnostic laboratories, or emergency department, the risk of contaminating the municipal trash with infectious waste was much lower. Along with the black bags in patient rooms, housekeeping staff at Baluster collected municipal trash from family lounges, the cafeteria, and administrative lounges, all of which had a low risk of generating infectious waste. Additionally, there was less risk posed by waste stream contamination because fewer staff members handled waste at Baluster, where the housekeepers who removed trash from patient rooms were also responsible for transferring the trash directly to the dumpsters outside. This increased the likelihood that they knew where the trash they were handling had come from and potentially increased their awareness of relative risks of the bags' possible contents.

Infectious Waste

What classifies waste as infectious? At both Hopewell and Baluster, any "dripable, pourable, or squeezable" materials saturated with blood or body fluids were considered infectious waste and therefore required disposal via red bags. Waste disposal policies considered a gauze pad soaked with blood infectious and therefore required it to be disposed of in the red bags. Meanwhile, gloves with a small amount of wet blood or covered in dried blood were not considered infectious and therefore belonged in the black municipal trash bags. Needles and other disposable devices that could potentially infect someone by piercing their skin were disposed of in sharps containers—small, red plastic boxes with lids marked with a biohazard symbol. On certain units where the frequency of creating infectious waste was higher, each patient room had its own red bag container. At the hospice unit, each patient room held a red bag container, though I never saw any dripable, pourable, or squeezable waste created during my observations at hospice. In contrast, Hopewell's conventional cancer and palliative care units provided red bins in the soiled utility room rather than patient rooms, though again I never observed the generation of infectious waste on the units. I did, however, observe nurses at Hopewell disposing of needles and other sharp materials in the appropriate receptacles. When full, red bags and sharps containers were collected and stored in the basement of each facility and then hauled away to the local infectious waste processor.

Since the problem of infectious waste entering the municipal trash seemed to me to be a potential public health problem, I was surprised to learn that there are more regulations to prevent trash and recyclables from entering the infectious waste stream than vice versa. A few decades ago, many hospitals had their own waste incinerators designed to handle infectious waste. The use of these incinerators to burn a variety of hospital waste led to two primary concerns: air pollution emissions and the unnecessary incineration of recyclable materials.[12] To reduce the amount of trash and recyclables entering hospital incinerators, many states began implementing regulations that required medical facilities to separate their various waste streams and prepare annual reports on infectious waste generation and reduction efforts.[13] These waste reduction and reporting regulations continue to guide waste handling.

During my time at Baluster Hospice, administrators determined that nursing staff members were throwing too many items into the red bags—items that did not qualify as infectious. As a result, the administration began a new training program for the nursing staff on infectious waste generation and handling. The program included hands-on training, reminders posted in strategic places, and an education campaign to increase awareness about the cost of incorrect disposal. During staff meetings, trainers would grab an infectious waste container and go through it item by item with the nurses, removing the things that had been improperly disposed of and putting them in the correct receptacle. The trainers had devised this education system because they had found that "visual evidence seems most helpful" for making staff aware of correct procedures. In addition, the trainers "reached each staff meeting with this information. It is tedious but important to reach as many people as possible and get the information into the minutes of the meetings to be read by staff not at the meeting." With the new training in place, the amount of infectious waste at Baluster Hospice decreased by over 40 percent, from about 1,300 pounds per year to around 560 pounds. Since disposal of infectious waste was around $650 per ton compared to $50 per ton of municipal trash, the facility reaped significant cost savings for diverting noninfectious materials from the red bag waste stream.

The high cost of infectious waste processing reflects intensive resource use in the disposal process. After individual hospital incinerators began shutting down, newer highly regulated incineration facilities began processing health care waste, and, until recently, these facilities incinerated over 90 percent of infectious waste in the United States.[14] The major ecological and public health concerns associated with these incinerators include emissions of mercury, particulate matter, hydrogen chloride, and dioxin. Since 2013, the Environmental Protection Agency has required alternative treatment and disposal of medical waste in order to decrease the risks associated with incineration.

Neither Hopewell nor Baluster disposed of infectious waste via incineration. Instead, they shipped their red bags to a local processing facility that collected infectious waste from medical facilities and institutions throughout the region. Originally designed as an incinerator in a joint effort of Hopewell and two other regional hospitals, in the 1990s the facility converted to using a series of steamers and microwave units to sterilize the waste. Infectious waste entering the plant was steamed to about 280°F, hydraulically lifted and dumped into a large grinder, and then sent up a conveyor belt through a series of six microwave units before finally being dumped into a trash compactor. At that point, the waste was deemed safe enough for the local municipal landfill. The annual total volume of infectious waste processed by the plant was about 2 million pounds (1,000 tons).

As might be expected, the personnel who handled the infectious waste stream had various concerns about potential exposures. Roger, an administrator at the waste processing facility, told me that the workers there often talked about the things that improperly ended up in the red bags. The staff was most attuned to large metal items like "big staplers or some mechanism, or even primarily coming out of surgery . . . they'll throw a stainless steel hip joint in with the red bags. Well, those sorts of things will plug up our machines." When the grinder locked up, he continued, "one of our guys, our maintenance guy or one of our operators has to go down inside of that hopper to either pull that part out or unplug it or we have to pull the shredder out. By doing that, you've exposed that person to all of those contaminants and hazards that are inside that hopper. You've exposed that person to all kinds of things."

Since the waste had gone through the 280°F steam treatment before reaching the hopper, it was safe according to guidelines. But Roger worried that maybe it had not been steamed for long enough, or maybe some unground sharps remained.

> So if our guys suspect that there's something in one of the red bags that shouldn't be in there, then they will open up the red bag [before it is processed] with the rationale that it's better to open it up outside of the machine and expose yourself to something that you have better control over. And you can put a face shield on and get more [personal protective equipment] on than getting yourself exposed to all of that stuff down in that hopper. So we say it's the lesser of two evils, and we choose to open that one bag up outside instead of waiting for it to get in there and plug up, and have to pull out thirty bags of stuff that's on top of that one thing.

Roger told me that these types of exposures were rare, but they did happen. When I asked what could be done at the hospital to reduce the risk to his crew, he told me,

> I think it boils down to training—making people aware of the hazards and familiar with the materials that they're handling, be it hazardous waste, or regulated medical waste, or body parts, or what have you. Whatever it is, making them more aware of why do you want to dispose of it this way, and by handling it this way you're protecting somebody downstream—get people to look at the whole system instead of just their own little closet. . . . And if you don't dispose of it the right way, somebody is going to have to intervene somewhere and pull it out of that box or bag or whatever and put it in the right category, and typically that's us. Well, and I shouldn't say it's just us, I'm sure some of the hospital people find that before it even gets to us, and somebody there takes corrective action. So I think it's a matter of training and making people aware of not just how and why to do it, but the implications that if we don't follow this there will be somebody else who is going to be exposed to a hazard, and we could eliminate that by just doing it right in the first place.

Single-Sort Recycling at the Hospital

When I first began observing at the hospital's conventional cancer and palliative care inpatient units, there were blue recycling bins in the administrative office, the nurses' station, the nurse computer area, and the nurses' break room. While many nurses used these containers to dispose of drink containers and such, whenever I scanned these containers they usually held more trash than recyclables. Although it appeared that recycling was not a major concern on the floor, oftentimes when I introduced my research interests to new staff members they would suggest that the units could do a better job with recycling in order to minimize their environmental impact. Eventually, the nurses talked about the idea enough that they decided to implement a recycling program throughout the units as one of their required annual quality improvement projects—including adding recycling bins to the patient rooms. Their plans gained momentum when they learned about the successful recycling program that a nurse had recently implemented in the operating room.

Jill was a nurse in the operating room who helped create the OR recycling program. When I asked why she had helped start the program, Jill responded, "Well, we just didn't have one! A lot of the stuff that could get recycled was just getting thrown in the garbage here." Jill received her OR training at another institution where recycling had been emphasized, so she knew that a lot of what was being trashed at Hopewell could instead be recycled. "With the amount of garbage we produce in a day in the OR—because everything is wrapped separately to maintain sterility—I was just kind of appalled at how much was going into the actual garbage. So that was pretty much why I wanted to start it."

After getting permission from her supervisors and buy-in from the housekeepers in the OR unit, Jill sat down with administrators from the Environmental Services Department and the company that hauls and disposes of Hopewell's waste and recycling streams to learn what could be recycled. She described how she had "brought a whole bag of stuff we use in the OR and we went through and learned which items are recyclable and which are not. And we learned what the rules are and how they do the recycling over at the center." Jill learned which materials were recyclable, and the processes that the materials would go through at the recycling facility. She developed relationships with people along the recycling chain, and she identified contacts at the facility who could help correct any problems.

For a week, Jill did a trial run of the new system with just her small operating team to iron out any problems that arose. One of the biggest obstacles the OR staff had to overcome was a concern about time. Appropriate recycling required separating the wrapper from each sterile surgical item into two pieces: half that was recyclable plastic, and half that was nonrecyclable Tyvek paper. Some OR nurses were concerned that peeling the two sides apart and separating them into the correct waste streams would take too much time, thereby "taking away from patient care because you couldn't get the patient in the room as fast," Jill said. "So I had a few people against that, they didn't want their time imposed on, but there were other people on the team who were all about saving the planet, and this extra second it took to separate was keeping [material] out of the landfill."

After ironing out some of the problems, Jill presented the new recycling program at a weekly OR staff meeting that included the entire staff of nurses and technicians. The relatively small staff size and the fact that the OR is not a 24/7 unit allowed Jill to share information about the project with all of the staff members who would be most intimately involved in carrying it out. Right after the meeting, she put a recycling bin and list of recyclable materials in each operating room and spent the day going room to room explaining the new system and answering questions, presenting herself as the go-to person who could help support and troubleshoot for the project.

At the same time, another OR nurse who had worked at Hopewell for a number of years teamed up with Jill to implement a program to decrease the amount of red bag waste coming out of the OR. Since the teams were throwing anything with blood into the red bags, even if it was not "dripable, pourable," there was a lot of waste entering the red bag stream that should have been in the municipal trash—and as a result the hospital was paying a lot more for disposal than it could have been paying. Since the OR creates about 27 percent of Hopewell's waste, the team anticipated that training the staff proper disposal procedures for the various waste streams could have a huge impact on the hospital's bottom line. In addition, Jill said that having the support of this well-known and well-respected nurse helped ensure the success of the recycling program by creating staff buy-in from people who may have otherwise been upset with the new procedures.

Despite the buy-in from the nursing staff and surgical technicians, one of the biggest obstacles Jill encountered with the program was that the surgeons contaminated the recycling bin with their used gloves. When I asked why she thought that was happening, she said, "They just have other things on their minds. I don't think they do it to be malicious, and I don't think they're doing it on purpose. They scrub out and they want to get out of there."

"And has there been education for the surgeons?" I asked.

"We've tried," Jill said, "but they don't have the big meeting like we do. They don't have the big team meeting because there are so many different residents coming through and med students. But the day that we started it, we did put up signs at all the scrub sinks so as the surgeons were scrubbing in they could read that there's this thing going on, and we aren't asking you to do anything, just be aware. That was all we really wanted from them."

"Do you have a sense that if a nurse sees the gloves going into the recycling bin, they grab them out?" I continued.

"Yes. Really, everyone has latched onto the whole thing, and some nurses are quite protective of their recycling bin—to the point that some nurses will take the bin and put it in a corner away from the other garbage and laundry so the surgeons don't accidentally throw something in there that they aren't supposed to!"

Despite the difficulties the program had faced, recycling in the OR was considered a success overall. After one year of monitoring the program, the staff determined that their work to divert noninfectious waste from the red bags and recyclables from the black bags decreased the costs of Hopewell's OR waste disposal from over $75,000 per year to about $17,500 per year. It seems that with such potential cost savings, administrators at Hopewell would be interested in implementing similar programs throughout the hospital. But in my observations, it was nursing staff throughout the hospital that spearheaded any attempts to increase recycling. Without broader administrative support in the form of training and funding for blue recycling bins, projects such as the one on the conventional and palliative care units were largely unsuccessful.

Nurses and housekeepers from the OR spread the news about the success of their new program, and staff members in other hospital units began contacting Jill to find out how they could implement change in their own areas. A few nurses on the conventional cancer and palliative care units learned about the program at a hospital nursing governance meeting and connected with Jill to learn more about what items could be recycled on the units. They discovered that a lot of the materials they were currently trashing could actually be recycled, including saline bags, IV tubing, and supply packaging. So they decided to implement a recycling program on the units.

To begin, the nurses in charge of the project shared information about their goal at the unit's monthly nursing staff meeting and then chose a day to place signs and recycling containers throughout the unit. One of the first challenges the project leaders encountered was a lack of space to place the new recycling bins. In the unit pharmacy room and supply closet they had to try several spatial configurations to get the new bins to fit without being in the way. In patient rooms, they decided that the best option was to replace a garbage bin with a recycling bin. Unfortunately they eventually discovered that even with an informational sign this bin placement would be a major barrier to the success of the program. The location was right next to the door exiting to the hallway, in a small cupboard with a swinging flap that provided access to the bin. This caused two challenges. First, it left the only other garbage can in each room under the sink, far from the door where clinical staff frequently entered and exited. Second, since a small recycling sign on the swinging cupboard door was the only information that announced the otherwise hidden blue bin, the project leaders quickly discovered trash in the recycling bins.

One of the biggest contaminants was gloves, and the nurses suspected that the biggest culprits were the physicians, though the housekeepers alleged that some of the nurses might be the responsible parties. Regardless of who was to blame, everyone reported that the likely cause of contamination was having the recycling bin in the spot where the trash had previously been located. As one nurse put it, "I think it's second nature because the containers are by the door and as you go out the door you throw your gloves away."

Another challenge to the program was that, compared to the OR, champions of the recycling program on the units could not guard the bins to protect them from contamination. In patient rooms, recycling bins were accessible to nursing staff (not all of whom had received recycling training), patients, visitors, physicians, and anyone else who happened to enter the room. Rather than being able to safeguard the recycling bins as had occurred in the operating room, the project leaders on the conventional and palliative care units had little control over who had access to the bins. The combination of these challenges soon led to suspension of the recycling program on the units. Administrators determined that too many nonrecyclable materials were entering the recycling stream and causing concerns at the Materials Recovery Facility (MRF) that processed Hopewell Hospital's single-stream recycling.

At the MRF, a complex system of people and machinery sorted and separated the glass, plastics, metals, and other materials. According to one staff member of the hauling company that transported the hospital's recycling, problems with Hopewell's inpatient recycling program arose when workers at the MRF began to see materials of concern passing through the mechanical sorting devices without being separated out, thereby requiring people to hand-sort these items from a conveyor belt. When workers on the line saw medical tubing, "they [would] think there must be a needle somewhere," and the whole process would come to a halt. Nursing staff on the conventional and palliative care units had been trained—correctly—that plastic tubing was recyclable. So their practice had been to place the tubing directly into the recycling bins after disconnecting it from the patient. Since Hopewell used needleless IV access, there were no needles dangling from tubing to begin with, and therefore no threat to the recycling facility workers. But the recycling staff did not know this and were alarmed at the possible threat of being stuck by a needle. The waste hauler suggested that the best way around such problems was for the hospital purchasing agents to work with suppliers to make sure that any materials that the hospital wants to recycle are made of the proper materials for mechanical sorting. This would ensure that these items would never make it into the hands of recycling workers, thereby increasing the potential for hospital recycling programs to succeed.

Recycling at Hospice

At Baluster Hospice, recycling bins were located in family lounges and at the nursing stations only. I asked Caroline, a staff member familiar with the various waste streams at Baluster, why there was no recycling program in the patient rooms. She responded, "We're sort of struggling with our entire recycling program here, which is more in the office areas than on the [inpatient unit], but part of it is that we just haven't been very successful in getting people to recycle and to understand that housekeeping is not going to pick the recyclables out of the trash for you. If you don't sort it yourself, it all goes in the trash." She added that clinical staff members on the unit were interested in training staff to recycle things like supply boxes and pill bottles, "So we are going to work on that from a staff perspective, but I'm not sure that we'll change what's going on in patient rooms and family areas."

"Is it an eventual goal to recycle things like oxygen tubing and other things that are recyclable but right now are just ending up in the trash?" I asked.

Caroline responded, "I would say yes from a staff perspective it would be a goal that anything that can be recycled is recycled. But if it's something that a patient or family member has in a room it is very hard to control.

People aren't always here for very long, it's not really something we want to focus on while they are here."

The general sense at the hospice facility was that patients and family members had more personally important things on their minds when obtaining care on their inpatient unit, and should not be bothered with concerns about recycling. However, if the system were designed to make recycling easy and efficient in patient rooms, the health of both the patient and the planet could be taken into account without need for dedicated effort.

Medical Supply Donations

Of all the problems related to waste at Hopewell Hospital and Baluster Hospice, the most pressing challenge I observed was the sheer volume of waste generated within these medical facilities. Fortunately, I also discovered people who were recovering "waste" materials and redistributing them as vital medical supplies. Scott, a volunteer and former nursing assistant at Baluster Hospice, had been working for a few years with a statewide nonprofit organization to collect and redistribute unused medical supplies to resource-poor areas around the world. As part of his task, he had been training Baluster's nursing staff to minimize the amount of materials they wasted. Before he helped launch the donation program in the hospice facility, a lot of unused materials had simply been thrown away—like in this chapter's opening scene on the palliative care unit with Alex—because infection control standards deemed any medical supply that crossed into a patient's room to be contaminated.

"But these foreign countries have no standards like that, and they can still use a lot of the stuff that we would throw away," Scott told me. He continued, "For a while we were just dumping everything." But when he learned about the nonprofit, he "saw the need for lots of this stuff we were just throwing away . . . anything from walkers to wash basins . . . you name it. Once it's been used, it can't be reused."

I noted that many of the nondurable items I had seen were also being wasted, even when they had been unopened. Materials like boxes of gauze or medical gloves were trashed, even if they had just been sitting on the counter.

"Right, but it's gone into the room," he said, and that means it was considered contaminated. "And you know, some of our patients are only here for an hour, and once you take it in [the room] you throw it."

Infection control is a major driver of resource use within health care. Nosocomial, or hospital acquired, infections affect 1.7 million hospital patients each year and lead to over 98,000 deaths annually.[15] It is in a hospital's best interest to prevent infections because minimizing the spread of pathogens improves patients' health outcomes and their satisfaction with care. Preventing infections also saves money. Hospital-acquired infections can take weeks of inpatient care to treat, with a cumulative cost of up to $45 billion per year for U.S. hospitals. Since Medicare and Medicaid rules suggest that certain things—like hospital-acquired infections—should never occur, these insurers will not reimburse medical facilities to treat such "preventable complications," which can cost up to $45,000 per patient.[16] So hospitals have worked to prevent these events from happening, and the reduction in nosocomial infections has benefited both patients physically, and medical institutions financially.

But in the case of clean, unused, and unopened supplies in inpatient rooms, the standards create a lot of waste of both money and materials. To overcome some of these problems, Scott and the Baluster nursing staff identified an opportunity to decrease the waste of resources created by infection control standards. By designating a certain area near the doorway of each patient room as a "clean area," any supplies left in that location were considered noninfectious and could be returned to the supply closet rather than thrown in the trash.

Thanks to Scott's efforts, Baluster Hospice had diverted nearly a ton of unused medical materials from the landfill each of the previous few years, both through restocking "clean" supplies and by donating "infected" supplies. Once or twice a week Scott checked in the designated areas throughout the facility where staff piled the donations, filled up his car with the supplies, and drove them to the nonprofit organization's warehouse where they were checked, weighed, sorted, and packaged along with donations from other regional health care facilities. From there, the materials were distributed globally to help fill requests from organizations in need. For example, Scott told me that in response to a recent earthquake event abroad, "between us and [a similar organization in a neighboring state] we sent down eleven semitrucks full of medical equipment." On a day-to-day basis, Scott suggested that donating individual items "seems like a little piddly bit, but in the long run it adds up."

Saving nearly a ton of unused materials from the landfill each year is an impressive accomplishment, but it provides further evidence that the implementation of infection control practices creates a large volume of waste that comes at a cost to health care facilities. One of Scott's goals was to increase awareness among the nursing staff that both the volume of waste and the cost of disposal could be decreased simply by bringing fewer supplies into a patient's room.[17] He explained that the nursing staff members "are getting better about being careful about what they take into the rooms because we're trying to cut down on cost. And there's a big amount of cost. Before, we would take a whole bag of [adult diapers] into the room and use one. The patient expires; you had to throw them—even though they were still in the package. It was just a crying shame what we were doing. So now the [nurses] just take in three or four [diapers] instead of the whole bag."

Information seems to be the key to the program's success. Before starting the program, Scott tells me, "I saw a lot of waste." But when he became aware of the donation organization and the global need for medical supplies, "I could see where we could use it," and the waste became a resource. The materials could either go to a landfill, or to communities around the world that needed them. Although the program has diverted a ton of materials from the landfill each year, Scott said that he hopes one day he will no longer have to pick up and transport the supplies for donation "because nothing will be wasted."

There are several factors that lead to the wasting of unused medical supplies. In the opening scene of this chapter, I observed Alex as he filled and discarded a 30-gallon trash bag of unused, unopened medical supplies on the hospital's palliative care unit. Earlier that morning as clinicians rushed about to care for their patient in her final moments of life, their focus was on responding to the patient's needs as quickly and efficiently as possible. It is likely that they brought extra supplies into the room in a hurry, not sure of what they would need to comfort the patient, and not wanting to delay care by needing to return to the supply closet to retrieve supplies. Given the disarray of the room, it is also likely that the packages had only been perched on a counter in the room for a short time. Since neither the nursing staff nor the housekeepers on the unit had been trained or were aware of the possibility to donate these items, the new, unused materials were landfilled.

And yet Hopewell did collect items to donate. Annually the hospital collected about 6,500 pounds of material, mainly surgical supplies for donation to the same organization that redistributed materials from Baluster

Hospice. Margaret, who worked in Hopewell's reprocessing unit—the place where reusable supplies were cleaned, sterilized, and packaged for reuse in the hospital—showed me a pile of donation items in the basement.

"Is there really something wrong with this stuff" I asked.

"You know," she said, "maybe it came from somebody's room who had an infection, I don't know. Maybe it was dropped on the floor, or the patient went home and [the billing department] couldn't credit them." She looked through the pile and pointed to a small, packaged item, "This one is out of date. It's all in plastic, there's nothing wrong with it."

I ask her why the system has evolved toward discarding these items, why they were not considered serviceable here, but were considered fine for donation. "It's all regulations," she responded. "It's the FDA and the government and our medical legal system and we have to protect ourselves." Margaret was echoing the need to protect the hospital from having to pay for hospital-acquired infections should a contaminated supply be passed along and infect another patient. But donating supplies that are considered contaminated by U.S. standards raises an ethical dilemma. Why are these materials not good enough for patients in our medical system, but considered fine for passing along to other countries?[18]

Several of the staff members I spoke with acknowledged this concern and responded that in areas of the world where medical supplies are scarce, perhaps having something is better than nothing. However, the donation of medical equipment to areas of the world that lack the resources to maintain or run the machines has led to veritable graveyards of broken equipment in the developing world. Estimates suggest that up to 80 percent of medical equipment in the developing world has been donated, and that only up to 30 percent of it becomes operational.[19] As the machines pile up, the ethical dilemma of medical supply donations is quickly becoming an environmental dilemma as broken machines are removed to dumping grounds, leaving impoverished communities both with our waste and without functional resources to provide health care to the local population.

Trade-Offs and Opportunities for Minimizing Waste

When I asked an Environmental Services administrator at the hospital how the environmental impacts of waste were considered in his job, he responded,

"I think as a hospital we try to minimize the amount of waste we produce . . . but you know, we're still a hospital, we're here to serve the patient. The patient comes first so it's not easy."

Proper disposal of each waste stream requires staff time and resources that could be spent on other tasks. The division of labor between various staff positions is, on one hand, a good thing; it means the staff is well-trained in their own area of expertise and therefore supports an efficient, high-quality system. Nurses do not need to know where the waste goes, and Environmental Services staff do not need to know the clinical care that produced the waste in order to do their own jobs well. But if nurses had knowledge about the waste disposal process, its financial cost, and the potential exposures to workers downstream, would they be more vigilant in ensuring the proper disposal of materials? If they knew the names of the housekeepers, infectious waste handlers, or landfill staff, would they be more likely to throw waste in the correct bag and know whom to ask if they had questions about proper disposal? The case of Jill's successful development of the OR recycling program suggests that such measures do make a difference while still maintaining high-quality patient care.

My observations suggest several opportunities for minimizing the negative trade-offs of health care waste streams without compromising a high standard for clinical care. One key example comes from my observations that Baluster Hospice had lower overall rates of waste. Within the hospice setting, costs of care drive the use of materials in ways that tend to limit waste. The Medicare hospice benefit provides a certain dollar amount per patient per day of care, which incentivizes hospice institutions to reduce their costs, and waste reduction is one key way to achieve this goal.

At the hospice facility, efforts to minimize waste included training nursing staff to take fewer items into a patient's room, and to leave supplies in the designated "clean area" where they could be restocked in the supply closet if unused. Several factors allowed Baluster to achieve such waste reduction, including the fact that both the facility and the staff size were relatively small. This made it possible to design a training program that efficiently relayed information about waste and the donation program to the entire nursing staff. In addition, the spacious room design provided space for the designated "clean" areas. Furthermore, the informal motto that "there is no emergency in hospice" likely helped decrease the number of supplies that were brought into patients' rooms during their final moments of life.

In contrast, Hopewell was a relatively large facility with a relatively large staff. While each nurse at Baluster Hospice could easily be trained and held responsible for removing donation items to a designated area, it could take a nurse or housekeeper on the conventional cancer and palliative care units at Hopewell Hospital twenty minutes to make a round trip to the basement to drop off donation materials. This suggests that a clear division of labor is necessary for larger health care facilities to design systems that minimize the wasting of unused supplies and returns us to the question of how to maintain social connection and obligation to protect workers downstream. There may be no clear answer that applies to all health care facilities, but I argue that the question is worth considering.

Likewise, the difference between the ability of Baluster Hospice and Hopewell Hospital to designate clean areas in patient rooms is a matter of scale. Patient rooms at the hospice facility were large enough to easily accommodate this design challenge and helped reduce waste generation and cost. In contrast, patient rooms at the hospital were relatively small and lacked space for easily designating a clean area. Given the contamination of the recycling program in patient rooms at the hospital, it is also important to consider that implementing a clean area strategy in patient rooms would require strict training protocols to eliminate the risk of contamination of unused materials. One nurse suggested that, given the design of the conventional care unit, a better plan would be to replace linen closets or cupboards in the hallway immediately outside patient rooms with designated clean areas to protect supplies from pathogens. While identifying ways to minimize the wasting of unused supplies via the use of designated clean areas has several potential benefits that reach beyond the walls of the hospital, strategies for accomplishing this goal must be specific to each facility.

Overall, the generation of waste within health care—especially the waste of large volumes of unused materials—and the ethical implications of donating materials to underserved communities when these supplies do not meet U.S. standards of infection control are offensive to the "Do No Harm" sensibility that underlies medical practice. These practices undermine the health not only of human and ecological communities downstream who bear the burden of disposal facilities but also of those living upstream where medical supplies originate. Therefore, chapter 3 turns upstream so we can better understand just what is being wasted.

3

Medical Supplies

Tracy, an experienced and knowledgeable nurse who splits her time caring for patients on both the conventional and the palliative care units at Hopewell, stopped in the unit supply closet to grab a handful of supplies before heading down the bone marrow transplant hallway to visit a patient. The charge nurse had assigned me to follow Tracy for the day, and as I explained my research to her, her eyes twinkled with an insider's knowledge. She lifted the newly acquired supplies and jingled them in front of me—some tubing, gloves, and a prepackaged dressing kit—and informed me to look into the procedural kits like the one she held.

The task at hand was to change a patient's IV tubing and wound dressing on his peripherally inserted central catheter (PICC) line. The PICC line is a long, thin plastic tube inserted into a vein in the arm and then guided up to a larger vein near the heart. PICC lines are one of the preferred IV access options for patients who need chemotherapy over long periods of time. As Tracy put it, in these patients, the "smaller veins are shot" from the repeated exposure to chemo drugs, so the PICC line works well because it conveys the pharmaceuticals into the larger veins. In general, IV tubing and dressings need to be replaced every three days to decrease the risk of infection. For patients with PICC lines, maintaining an uninfected IV line is particularly important since the interior end of the line is very near the patient's heart, and any infection could quickly become life threatening.

When we reached the patient's room, Tracy put on her blue nitrile medical gloves, removed the outer plastic packaging of the dressing kit, and created a sterile plane on the bedside table by unfolding the inner wrapping of synthetic blue paper to reveal the kit's contents. Inside were four different types of alcohol wipes and swabs, an iodine stick, a sticky foam dressing that would rest against the patient's skin, another dressing with a clear plastic window that would affix to the sticky foam and form an outer protection layer, a "stat lock" stabilization device to tape the IV lines down to the patient's arm, a sticker to write the date of the dressing change, and a pair of vinyl gloves.

"It's such a waste! We never use a lot of this stuff," Tracy told me after she had finished changing the dressing. She folded up the packaging along with the unused items—half of the alcohol wipes and swabs, along with the iodine stick, stat lock device, date sticker, and vinyl gloves—and tossed them in the garbage.

A few months later while I was interviewing a medical supply administrator at Hopewell, I asked him about the kits. "Lots of vendors make them," he told me, because manufacturing sterile kits saves companies money. Rather than having to package and sterilize each item individually, kits are a way to bundle a number of items together and sterilize them all at once. "Overall, kits cut down on packaging and processing and they save nursing time since the nurse just has to grab one item and go. The danger is that you have to make sure you configure the packs with all the things that will be used." Opening a kit compromises the sterility of all its contents, so any unused materials must be discarded. "So if you put too many materials into a pack," he said, "it's very wasteful."

"So do you as the users get to decide at all what is in those kits and packs?" I asked. "Because I saw that a lot of the materials are being wasted in some kits."

"Yes and no," he replied. "If it's a standard kit that is sold to lots of hospitals, those are cheaper but there's a higher chance that you aren't going to use everything in that kit. Or you'll have to add stuff to that kit if it's not configured to how you do things here. You can configure your own kits, but that would be custom, and those cost more money because the manufacturer has to go outside a normal bulk process and custom make those for us. So there are pluses and minuses both ways." In the case of the PICC line dressing kits, the trade-offs had swung toward cost savings, rather than waste reduction. But what exactly are we wasting, and who are we affecting both upstream and downstream of the patient's bedside?

The Ecological Costs of Medical Supply Chains

If clinical staff knew that the production, manufacturing, and transport of medical supplies were contributing to environmental degradation and leading to health impacts around the globe, would they be more conscious in their procurement, use, and disposal of these materials? That, in effect, is the question posed by one of the few environmental analyses of a medical supply chain.[1] The analysis, completed in 1999 by Jessica Pierce and Christina Kerby, was a thought experiment that asked readers to consider the ethical conflict between individually based medicine and the health of the communities—both human and ecological—that produced the materials used in health care. Pierce and Kerby employed what they called an "ethical life-cycle analysis" to describe the potential harms resulting from the manufacture and distribution of latex gloves. Their methodology built on traditional life-cycle analysis, which is a method for assessing the environmental impacts of a supply chain beginning with natural resource extraction (the "cradle" or beginning of the material), all the way through processing, manufacturing, transport, distribution, and final disposal of the material (or the item's "grave").

At the time, Pierce and Kerby estimated that medical care in the United States used nearly 13 billion latex gloves annually, making them one of the most ubiquitous medical supplies. At first glance, the gloves' simplicity made their potential environmental impacts seem relatively benign. However, as the researchers uncovered various aspects of the movement of latex gloves from cradle to grave, they discovered a range of environmental and human health concerns.

Latex originates as a milky sap from rubber trees of the genus *Hevea* and is itself considered a renewable resource. However, in order to produce the large volumes of latex desired for commercial production of materials, including latex gloves, large swaths of native rainforest habitats were cut down to make room for *Hevea* tree plantations in several regions of the world. Several countries in Southeast Asia were among the largest producers of latex, including Malaysia, Thailand, and Indonesia. Pierce and Kerby contended that the destruction of native rainforest habitats to make room for latex plantations had led to soil erosion, decreased water quality, and loss of biodiversity. In turn, these environmental impacts led to human health impacts for nearby communities who experienced decreased yields of subsistence crops from soil erosion, impaired drinking water quality from siltation, and food insecurity from the loss of biodiversity.

Further along the medical supply chain, latex was transported to glove production facilities, many of which were located in low-income countries where companies could take advantage of cheap labor and nonexistent or unenforced environmental and occupational health regulations. The authors cited research from the time providing evidence of the poor working conditions and lack of access to clean water, sanitation, education, and health care in the communities that manufactured latex gloves.

Next the authors turned to the clinic itself, detailing the movement of the finished gloves from the loading dock where they arrived in boxes after being hauled long distances in carbon-emitting transport, to the final moment of brief medical use. Pierce and Kerby suggested that medical care required the gloves for perhaps two minutes before they entered the waste stream. Those gloves that were destined for a landfill would eventually decompose, though in the process they would contribute to the "landfill crisis"—too much garbage and not enough landfill space. In comparison, gloves discarded as infectious waste were destined for the incinerator, where they would burn at temperatures reaching at least 1800°F. The resulting fly ash (fine particulate matter released to the smokestack) and bottom ash (unburned solids destined for a hazardous waste landfill) contained toxins such as lead, mercury, chromium, and cadmium. These heavy metals posed health risks to workers and communities downwind of the smokestack or downstream of the landfill, where leachate infiltrated groundwater.

Pierce and Kerby's analysis showed us that the environmental and community health impacts of latex gloves were not restricted to the rainforests of Southeast Asia. They were also right here in our own backyard. Given the cradle-to-grave nature of the materials we use to treat patients, the clinical practices employed within our local hospitals have ecological costs that reverberate globally. While chapter 2 focused on the unintended consequences of waste as one of the most visible aspects of patient care in my three research sites, here we turn to look upstream at the environmental flows of medical supplies that help support patient care. Through this analysis we will begin to develop a sense of the broad impacts of resource use in my research settings, and the factors that govern supply use. We will also begin to see that in many cases, it is nearly impossible to uncover the complete story of a medical supply's journey. To get a better sense of the barriers preventing this broader view, we begin our analysis in the supply closet.

The Supply Closet

Cramped, full, and overflowing, the staff at Hopewell stocked and restocked the supply closet shared by the inpatient cancer and palliative care units twice a day. After a nurse buzzed me into the closet with her access card, I stepped into the crowded room to record the material culture of the hospital. Before I could begin my task, the door opened behind me, and a nurse entered to retrieve a bag of saline. When she exited the closet, I began taking note of any information that could help me track the flow of medical supplies.

My first notes were of bagged intravenous solutions. There were multiple concentrations of saline, dextrose, and potassium chloride, all ranging in size from 250 to 1,000 milliliters—and each bag was contained within plastic packaging. Next were boxes of latex and nitrile gloves in all sizes. Face masks, urinals, pitchers, tissue boxes, and plastic specimen collectors were stacked on the bottom shelves. My paper was filling up quickly, and I had only just begun.

The door opened again, and two nurses entered the closet in search of a kit to replace an IV port on a patient. I stepped into the hallway to give them room to maneuver the sliding shelves in their search, and as they left, I took the place they vacated to begin recording again. A few shelves later, I discovered an arm board used to stabilize a patient's arm when starting an IV line and made a big note: "reusable!" I underlined it twice. Until this moment, everything else I had recorded had been disposable after a single use. After completing my task a few hours later, the arm board remained one of the very few reusable materials I had encountered. Overall, 123 (91 percent) of the 135 total items stocked in the supply closet of the conventional and palliative care units were disposable

As I continued my exploration, some labels told me that products contained latex—an allergen that many hospitals have tried phasing out because of adverse reactions in both patients and staff. Others indicated that they contained di(2-ethylhexyl) phthalate (DEHP)—a chemical that makes plastics flexible and is associated with various health risks to vulnerable patients. Several items listed Canada, China, Costa Rica, Denmark, Israel, Japan, Malaysia, Mexico, or Singapore as their point of origin, but many supplies had no information about their origin. In noting this lack of information, I began wondering about all the places and hands these materials had passed through during their production and distribution. No pictures adorned the boxes to identify the landscapes where the raw materials had originated. No text listed the names of workers who had manufactured the products. There were no labels describing the distances the items had traveled, their ecological footprints, or the total cost of their cradle-to-grave life cycles. These details had never even made it to the loading dock.

The same was true for the materials stocked in Baluster Hospice's supply closet, and many of the materials were identical to the ones I had recorded at Hopewell. I began, just as I had at Hopewell, with the shelves of IV bags. Again, they were plentiful and came in various volumes and concentrations of saline. A plethora of bandages, wound dressings, and gauze products filled another few shelves, many from China, some from Thailand, and a few from Sweden. Several respiratory products were labeled as being made in Mexico, including oxygen masks, nebulizers, and oxygen tubing. I noted and underlined "made in the U.S.A." from the labels on mouth swabs and alcohol-free mouthwash. Overall, the list of supplies that I cataloged at the hospice facility was similar to that from the supply closet at the hospital, though the closet itself was more spacious and experienced less foot traffic since the hospice nurses tended to visit the room only when they needed to restock the smaller supply cabinets on each patient pod.

It is not surprising that in my observations of patient care, some of the materials I had recorded in the supply closets were used every day, whereas others were required only occasionally or in emergencies. The most commonly used items differed among my three research sites. On the conventional cancer unit, the most used supplies in order of volume included medical gloves, IV pumps, IV bags, IV tubing, alcohol wipes, hand sanitizer, and plastic pill cups. On the palliative care unit, medical gloves, hand sanitizer, IV pumps, IV bags, IV tubing, alcohol wipes, and mouth swabs topped the list. In comparison, at Baluster Hospice's acute care inpatient unit, medical gloves, handwashing soap and paper towels, hand sanitizer, mouth swabs, alcohol wipes, wound dressings, incontinence briefs, and catheters topped the list.

To triangulate my notes from the supply closets and verify my observations of materials used in delivering patient care within each setting, I obtained annual supply use data from both Hopewell and Baluster (see table

C.1 in Appendix C). Supported by these combined observational and institutional data, I identified medical gloves and polyvinyl chloride (PVC) plastic as the two most commonly used materials across all three of my research sites. Building on Pierce and Kerby's ethical life cycle analysis of latex gloves, I decided to follow the life cycles of these materials from cradle to grave to better understand the unintended consequences of medical care today. Medical gloves would again be the starting point, but this time, instead of creamy white latex, I would be seeking the story of blue and purple nitrile.

The Story of Nitrile

Now, over two decades after Pierce and Kerby's ethical life cycle analysis showed how the latex glove supply chain was harming environments and health on a global scale, latex has mainly been phased out of many U.S. medical settings. Concerns over the environmental and community impacts of latex were not the driver in this switch. Latex gloves have been replaced by a synthetic nitrile alternative because of rising concerns over latex allergies and higher costs associated with the latex supply chain.

Hopewell Hospital was in the process of transitioning fully to nitrile gloves while I was observing on the conventional cancer and palliative care units. Perry, a medical supply administrator for the hospital, leaned over to pull a copy of the facility's annual glove usage report from the printer for me. He told me, "We've been wanting to [switch to nitrile] for a while due to latex allergies, but what is actually pushing us this time is that the cost of latex is going through the roof. The cost of latex now is very comparable to the cost of nitrile."

According to the printout, Hopewell used a total of almost 13 million gloves (about 6.5 million pairs) each year. The conventional cancer and palliative care units' glove use accounted for about 5 percent of the hospital total. The hospital's annual cost of gloves was about $536,000. Latex gloves cost about 3 cents each, while the nitrile version was just under 5 cents for each glove. But this was changing. The cost of latex was increasing globally, and latex gloves were about to jump in price to just over 5 cents per glove. If the hospital continued to purchase the same number of latex and nitrile gloves as in the past, the annual cost would increase to about $637,000 per year. By switching to all nitrile, the cost would only increase to $621,000 annually. Perry estimated that by switching to all nitrile gloves to avoid the increase in latex costs, the hospital would save almost $16,000 per year.

I asked Perry if he knew why latex prices were skyrocketing. He responded, "Basically, all the latex gloves are made at this one factory in—I'm not sure if it's China, or Indonesia, or wherever—but latex being a natural item has to get shipped to this place. And I guess the cost of latex is up and the cost of shipping is up, and therefore it just pushes the cost of latex gloves up." The cost of nitrile gloves, on the other hand, was remaining steady for the time being.

After I had asked further questions about medical gloves, Perry put me in touch with the sales representatives he worked with at the nitrile glove manufacturing company. From this referral, I was ushered along to a technical representative for the company who taught me that this synthetic rubber consists of two main chemicals (acrylonitrile and 1,3-butadiene), which together create a material impermeable to liquids. Because of this impermeability, nitrile gloves had become the "gold standard" for protecting health care workers from pathogens and chemicals—particularly those used for chemotherapy. Since they were designed for disposal after a single use, they were also key at helping minimize the spread of infectious pathogens between patients. So they certainly protect the health of people within health care settings. To find out how they might affect the health of workers in the supply chain, I had to follow them upstream.

All of the company's gloves were made in Malaysia, where, as the company reps told me, the costs, processes, and quality were watched closely to maximize profit. The manufacturing itself required mixing together a number of chemicals purchased in volumes ranging from 55 gallon drums to full tanker trucks. When the nitrile chemical compound was mixed and ready, it was poured into vats at the "dip-line." Thousands of porcelain forms in the shape of outstretched hands hung from an automated conveyor belt encircling the room. These porcelain hands danced along the dip-line, dipping first into cleaning solutions, then into the nitrile compound itself. Depending on the specific nitrile formula needed for filling a particular order, the hand would emerge with either a blue or purple glove. The conveyor belt then ushered these color-coated hands toward dip-tanks of water and other chemical soaks, then on to driers. At the end of the line, the finished gloves were removed from the dancing hands, completing their ballet in as little as fifteen minutes. Depending on their final use and destination, some of the gloves were ready to undergo quality checks and packaging as soon as they were removed from the dip-line. Others required further processing, for example filling powder-free gloves with chlorine gas to remove any chemical powders and residues.

While my contacts at the manufacturing company were very informative about the technical glove-making process, they were U.S. based and, never having traveled to the Malaysian plants, were unable to answer my inquiries about the people and places that produced the gloves. Nor could they tell me the origins of the raw nitrile materials, or the impacts that resource extraction may have had on the local environment, workers, and communities in those locations. This information stayed behind as the 55-gallon drums and tanker trucks filled with local products were hauled away. Since nitrile gloves are produced abroad with environmental and health regulations and enforcement that are likely limited, information about the health outcomes of workers and communities stays behind as well. To find answers to the question of how nitrile production affected the environment and health, I had to follow in the footsteps of the researchers who had asked these same questions of latex gloves all those years earlier: I had to turn to the literature.

To begin, it is important to note that nitrile rubber is commonly used in the automotive and aeronautics industries, so health care is certainly not solely responsible for the cumulative environmental, occupational, and public health impacts that result from nitrile production. However, the fact that the manufacturing process of nitrile gloves poses environmental and health hazards is indicative of the tension that exists between medical care at the individual level and the health of the global ecological community. Keeping this in mind, I decided that the most obvious place to begin understanding these broader impacts was to follow the two main ingredients of nitrile: acrylonitrile and 1,3-butadiene.

At the very basic level, the two primary chemicals needed for the production of nitrile are derived from petroleum. As such, they originate somewhere in the global network of oil fields and offshore drilling rigs. If the petroleum was drilled in one of the top petroleum-producing countries—the United States, Saudi Arabia, Russia, Iraq, Iran—the environmental and human health implications of nitrile gloves must include air and water pollution, habitat fragmentation, agricultural land degradation, and climate change resulting from both oil drilling itself and the military-industrial complex that keeps petroleum flowing from these nations.[2] If instead the petroleum originated from an offshore drilling rig in Malaysia itself, we would need to consider the risks posed to marine ecosystems as a primary environmental impact of nitrile glove manufacturing. Malaysia is located in one of the most biodiverse regions of the world and is estimated to be home to 20 percent of all animal species found on Earth, including numerous species found nowhere else on the planet. Marine biodiversity in Malaysian coral reefs, seagrass beds, and coastal mangrove forests provides a number of ecosystem services that support human health and well-being.[3] In addition, marine ecosystems sequester more carbon every year than terrestrial forests, helping to mitigate the effects of climate change.

Oil spills, production discharges, and habitat disturbance from offshore drilling all threaten the marine ecosystems that benefit the health and well-being of Malaysian residents. During the period from 2014 to 2016, the Malaysian Department of Environment recorded fifty-one oil spill incidents, including both Tier 1 (minor) and Tier 2 (medium-severity) events.[4] A lack of baseline data on marine biodiversity in the area makes it impossible to know the full ramifications of oil spills in the region, but even low-level oil spills release pollutants that wash through the mangrove forests and coral reefs of the region, affecting fisheries, tourism, and the health of people and native species alike. Again, the production of nitrile gloves for use in health care accounts for only a portion of these impacts, and yet these connections are important if we wish to foster a broader understanding of the unintended consequences that result from resource use decisions and practices within health care. Consequences that affect both ecosystems where raw materials are harvested and people who work within the industry.

Workers within the synthetic rubber industry face a number of health risks.[5] Occupational health and safety are major concerns since long-term exposure to chemicals in the process is linked to cancers of the bladder, blood, lung, and stomach. The World Health Organization International Agency for Research on Cancer lists acrylonitrile as a possible carcinogen, and 1,3-butadiene as a probable carcinogen. Workers exposed to acrylonitrile—which is converted to cyanide within the body—may experience eye irritation, respiratory distress, and death as the chemical causes problems with the central nervous system. Likewise, exposure to 1,3-butadiene can cause eye, nose, and throat irritation and damage to the central nervous system, heart, and lungs, and it has been associated with increased incidence of leukemia in synthetic rubber workers.

Within the nitrile glove industry specifically, a report from the U.S. International Trade Commission during the COVID-19 pandemic raises human rights concerns for the workers who produce medical gloves.[6] The report stated that in July 2020 and again in March 2021, a Malaysian corporation which produces one quarter of the global supply of medical gloves was found to be using forced labor within its supply chain. As a result, U.S. Customs and Border Protection seized all nitrile gloves entering the United States from the responsible parties during inspections. Of interest is that the report focused on the effect that these seizures had on nitrile glove supply and costs, while also stressing the need to identify alternative glove producers, especially to meet the increased demand for gloves during the pandemic. Importantly, the document suggested that COVID-19 vaccinations in the United States alone would require 660 million gloves since infection control standards required a new pair of gloves for each vaccine administered.

By the time each glove has made its way from oil field to tanker, then danced its way along the dip-line, into the cargo ship, across the globe, through the port of entry, onto the long-hauler truck, through the distribution facility, onto the hospital loading dock, up to the supply closet, and onto the hand of a doctor or nurse, it has contributed to cumulative impacts to people and landscapes along its journey. All to provide patient care for an average of two minutes. And then the glove is thrown into the trash, where it will make its way to the landfill or incinerator to complete its cradle-to-grave journey. Pierce and Kerby asked whether clinical staff would be more considerate when using medical supplies like gloves if they were aware of these impacts. However, in my observations and interviews it became clear that the unintended consequences of medical supply chains require system-level change rather than increased awareness among clinicians. A more effective question might then be whether clinicians could advocate for change.

Medical Gloves: A Box per Day

Liz, a nurse who divided her time between the conventional cancer and palliative care units, said she was not surprised that workers would become ill because of all the chemical exposures, then added, "But do you know if there are alternatives? Are latex gloves better?"

"Well," I replied, "they have different health impacts. Consider rubber tree plantations in Southeast Asia. The plantations cause soil erosion and water pollution in that localized area, and that has health effects on the people there."

"Yeah, I can totally see that," Liz responded. "So I don't really know what you would do. I mean yeah, you're right. It's ironic that we're making some people sick to make other people better, but it's hard to say that a person doesn't deserve that treatment. Though, I don't know, is it right for us to give [second time] bone marrow transplants for people when it didn't work the first time? I mean, there's no evidence that it will work the second time, so why do we do that? But it is being offered to patients."

Administrators in the purchasing department at Hopewell echoed this sentiment when I asked whether information about the health of manufacturing workers would be considered in purchasing decisions. One administrator replied, "I think that would be driven by the users because if the items came with warnings or if they heard through the media that the use of this product is causing [illness], or some poor kid in a sweatshop in India is making this, yeah I think it would [be considered]." After a moment he added, "Then again you'd have to see, is there something else? Let's say the alternative is better in that respect but it won't do the same job for the patient. So are you willing to say that you would go and do something less for the patient because of this concern about something else?"

Medical gloves serve two purposes. The first is to protect the patient, and the second is to protect the clinician. Monica, a nurse who I observed on the conventional cancer unit, explained that "anytime you come into contact with body fluids really is when they protect us," she said. "Urine. Stool. Stuff like that. Just touching the patients, most of the time on our unit, it's to protect them. You're washing your hands then putting on gloves to protect them from spreading germs from other patients or from yourself."

As discussed in the previous chapter on waste, infection control standards are complex safeguards mandated to ensure the safety and protection of patients.[7] Gloves are one of the primary defenses against hospital-acquired infections. And the nurses at Hopewell go through a lot of gloves. Liz told me, "I go through a box a day." She knew because she was allergic to latex and set aside boxes of nitrile gloves to make sure she would always have them available in her size. Likewise, when I asked Monica how many gloves she used in a day, she replied, "Oh god! Every time you walk into a room on our unit we use them. You're supposed to use them every time you have contact with a patient, and sometimes, you know, if I empty someone's toilet I'm going to change gloves. . . . I mean you go into every patient's room an hour. Four patients times eight hours is thirty-two, and then I'd say more than that. I'd say easily fifty pairs in a day,"—the equivalent of one full box.

If each one of those gloves were used to fulfill its purpose in protecting patients and staff from possible exposures the question of individual medical care versus community impacts would become grayer. But in my observations, a fairly large proportion of gloves were tossed into the garbage without ever actually coming into contact with a patient. The nursing staff at Hopewell are trained to grab a pair of gloves before entering a patient's room so they are prepared to provide any care needed. Only on occasions where the nurse knows ahead of time that they will not need gloves do they enter a room barehanded—for example, when a patient's IV pump alarm is beeping and the nurse just needs to reset the machine. But oftentimes after entering a room wearing gloves, a nurse would decide that she needed another supply before actually making contact with the patient, and as she exited the room to go to the supply closet, the unused gloves were thrown in the trash.

Dr. Smith, a palliative care physician, took the waste of gloves a step further:

> I think that medicine can be extremely wasteful. Just look at gloves and how many times we walk in and out of a patient's room and get a new pair of gloves. When I first started, maybe I shouldn't admit this, but when I first started, if you knew you were going to go back into the patient's room and you had gloves on, you put the gloves in your pocket and put them back on when you went back in. Now we have other things to worry about. We don't want to spread MRSA [methicillin-resistant *Staphylococcus aureus*] or anything we don't know, and so we throw the gloves away and walk away, and then come back and put on another pair of gloves.

The invisibility of infectious agents, the institutional pressures to minimize hospital-acquired infections, and the governance decisions that prioritize the control of such infections over waste are driving the use—and waste—of gloves.

Unintended Consequences of PVC Plastics in Medical Care

Just as in the case of nitrile gloves, the plastics used in medical supply chains are born of petroleum, and they share the same life-cycle impacts of oil extraction and transport already discussed for nitrile. But different types of plastics pose different types of risks because of their chemical makeup. Of all the plastics I observed in use at my three research sites, PVC appears to pose the greatest risks. PVC was used at both Hopewell Hospital and Baluster Hospice in flexible tubing for IVs, subcutaneous infusion sets, oxygen, and catheters, as well as in plastic bags containing IV solutions and pharmaceuticals. The primary environmental and public health concerns about PVC focus on the use of mercury, dioxins, DEHP, and other highly toxic persistent organochlorines.[8] The release of these chemicals impairs the health of the environment, manufacturing workers, and communities living near industrial production plants, as well as those who live near incinerators.

Mercury, a toxic element, is released into the air during the plastic manufacturing process and is eventually deposited in surface waters. Once it enters lakes, rivers, and streams, the metal reacts with naturally occurring chemicals in the environment and is converted to the highly toxic compound methylmercury. In this form, it is easily absorbed into aquatic organisms and then bioaccumulates. This absorbed mercury poses health risks throughout the food chain, and especially to top predators—including humans. As a result, local, state, and federal governments advise people to limit their consumption of certain fish species that are most likely to contain high levels of mercury because this contaminant can affect reproduction, growth, neurological development, and behavior.

Dioxins are a family of chemicals that can be released through forest fires and volcanic eruptions, but their main sources are industrial processes, including plastic manufacturing and waste incineration. In 2001, the Stockholm Convention on Persistent Organic Pollutants—an international treaty to minimize the production and use of toxic chemicals—listed dioxins among the "dirty dozen" chemicals of concern because of their ability to persist in the environment and bioaccumulate through the food chain. These chemicals are highly toxic and listed as "known human carcinogens" by the International Agency for Research on Cancer. When released into the air through industrial processes and waste incineration, dioxins can directly harm workers as they breathe the contaminated air, but they are also deposited on the land and water where, like mercury, they can contaminate food. They are now so commonplace in the environment, the World Health Organization states that all humans have some level of dioxins in their bodies, though our understanding of the health effects of dioxins has come mainly from large-scale contamination incidents, including an industrial accident that occurred in Seveso, Italy, in 1976. The chemical plant, owned by the Swiss multinational pharmaceutical company Hoffman-La Roche, experienced a mechanical failure and accidentally released dioxin into the atmosphere. Within a week, plants and animals in the exposed region began to die, and children were hospitalized for skin lesions. The community was evacuated, and because dioxin is particularly harmful to developing fetuses, some exposed women chose to abort their pregnancies. To prevent contamination of the food supply, more than 70,000 exposed livestock were slaughtered. Long-term epidemiological studies of the exposed population show an increase in cancers of the lung, colon, and blood (lymphoma and leukemia), as well as diabetes, respiratory disease, and cardiovascular illness. Notably, the community that suffered this exposure did not know that the industrial plant in their midst posed any health risks until after the accident.

For those who work directly in the plastics industry, the health effects of the vinyl chloride monomer—the building block of PVC—is better known, and occupational health and safety regulations can help minimize harms.[9] Since the 1970s, evidence of the health impacts of vinyl chloride exposure on worker health has been mounting. An early cluster of vinyl workers with a rare liver cancer led to decades of research on the health effects of this chemical compound, and we now know that vinyl chloride causes angiosarcoma of the liver. For the communities who live downstream and downwind from plastic manufacturing industries, chemical pollutants leaching into groundwater and filling the air can lead to increased rates of cancer as well and have earned the industrialized Mississippi River corridor of Louisiana the title Cancer Alley. Increasing evidence about the risks of vinyl chloride pollution led to many regulations and occupational safety and health protocols to limit worker exposures in the United States, not to mention the fact that the Louisiana towns of Morrisonville and Reveilletown were bought out and moved to reduce residents' exposure to toxic releases from nearby plants.

But, as I remember the labels on many of the products in the supply closets, most of the PVC-containing medical supplies had not been made in the United States, where regulations and enforcement measures are in place. Given the difficulties in regulating and enforcing environmental and health and safety standards in our own country, and concerns in developing countries that such regulations lead to market failures, there is limited evidence that worker, community, and environmental health are being better protected by manufacturers abroad.

While mercury, dioxins, and vinyl chloride pose serious health threats through their cradle-to-grave life cycles, the risks posed by DEHP are perhaps the most insidious because they can directly harm the most vulnerable of people through medical care itself.[10] DEHP is a "plasticizer"—a chemical that makes plastics flexible—and is used in many applications, ranging from shower curtains to children's toys, rain jackets to garden hoses. It is also commonly used in medical supplies, including IV bags and tubing. When used in IVs, DEHP can leach out of the plastic and enter a patient's body directly through the IV line. DEHP exposure can damage the liver, lungs, kidneys, and reproductive system, and is associated with neurological impairment, lower IQ, hyperactivity, and attention deficit. Longer-term exposure to DEHP in IV tubing among male infants in the neonatal intensive care unit can be particularly harmful to developing testes, and alternative DEPH-free materials are preferred in these settings.

Of course, given the many uses of PVC in modern society, medical supplies make up only a fraction of the global demand for the material. As a result, it may seem that health care institutions can do little to effect change in the petrochemical industry. However, since the missions of health care institutions often revolve around promoting and enhancing the health and well-being not only of their patients but of their communities, supporting alternatives to PVC and other products can promote these institutions as leaders in social justice and environmental and health equity. Fortunately, clinically acceptable alternatives to PVC are available for several medical supplies, though they tend to be more expensive than the PVC versions.[11] The choice of which option to select tends to be made within institutional purchasing departments.

Product Evaluation

Cultural factors certainly play a role in the use of medical supplies, but other governance issues play a role in the selection of items that make their way into the supply closets in the first place. Several institutional governance factors became apparent while I was working alongside Sean, a staff member in Hopewell's purchasing department. Sean spends most of the day clicking through screens on his computer, approving supply orders for various units at Hopewell, but he is also charged with meeting company representatives who are trying to persuade Hopewell to purchase their products. To streamline product evaluation, Hopewell advises these salespeople to meet with purchasing agents rather than with clinical staff. In this way, the purchasing agents work as middlemen to learn about products from the salespeople, research the cost and clinical evaluation of items that may be beneficial at Hopewell, and then present the items at a monthly meeting of the institution's Product Evaluation and Standardization Council. On this particular day, the sales rep was trying to sell Sean on the idea of purchasing a new valve to use with IV lines.

The rep's primary selling points included cost and infection control. The new valves, which could be reused multiple times, would replace Hopewell's current standard product, a small blue cap that helped maintain sterility by covering the end of an IV line every time the line was disconnected. The current single-use caps cost 60 cents each, and the new multiuse valves cost $1.12 each. Other hospitals had done research on the new valve and determined that they helped decrease infection rates. This independent research was an important aspect for Sean because it paralleled the Council's interest in choosing products according to evidence-based results.

After the rep finished making his pitch, Sean proceeded to explain a bit more about the financial benefits of infection control. He raised the issue of Medicare no longer covering the cost of care for hospital-acquired infections, and he provided an example of the steps Hopewell was taking to minimize infection rates, which would in turn increase revenue. He told me that, for example, if Medicare reimburses the hospital $10,000 for a typical knee transplant, but the actual procedure and follow-up care costs only $6,000, the hospital comes out $4,000 ahead. However, if the patient acquires a nosocomial infection after surgery, at a cost of $5,000, the hospital ends up $1,000 in the red because Medicare did not pay for the infection-related care. Since this new product had the potential to reduce hospital-acquired infections by keeping IV lines sterile, it would be worth researching the item further to bring to the Council. If there was enough evidence that the new product could be a good option for Hopewell, the Council—consisting of nursing staff and administrators, purchasing agents, infection control and risk management specialists, and supply administrators—would then review the item's quality, safety, standardization, cost, and potential impact on nursing practice, as well as Hopewell's contract status with the manufacturer of the product that would be replaced.

If all indications were favorable, the Council could decide to trial the item or make an outright switch to the new product.

I asked Sean if environmental factors ever arose, either in sales pitches or during product selections at Council meetings. He replied that he had not encountered any environmental arguments in his work, but my question made him curious. He swung around in his chair and brought up the website of the main group purchasing organization that Hopewell contracts with for supplies. After a quick search on the site, he found a list of "environmentally preferred purchasing"[12] options and over 300 contracts for environmentally preferred products from other members of the group purchasing organization. Sean told me he was surprised he had never seen this page before. Meanwhile, I decided I should get permission from nursing administration to observe a few Council meetings to better understand whether environmental factors ever arose in product selection. And indeed, environmental factors never did arise in the two meetings I attended. But I did learn that all the products discussed were disposable.

I asked one of Hopewell's supply administrators who works with the Council about the hospital's use of disposable materials and whether the environmental costs of disposables ever enters the product selection discussion. He told me, "There are times when we decide not to switch products because of environmental concerns. For instance, a lot of operating rooms use disposable gowns and drapes and other things. We actually use a lot of reusable drapes and gowns because we're one of the owners of a laundry service here in town, so it's less expensive for us to use those. But every year disposables manufacturers come to us and say they could save us a lot of money if we went to disposables and we elect not to do that." He paused briefly, then added, "Having said that, we still do use a lot of disposables because there just isn't a reusable version that works as well." This statement helps to explain why the majority of medical supplies I observed in use were disposable. But this had not always been the case at Hopewell and Baluster.

Reuse and Reprocessing

Not long ago, practically every medical supply was reusable. All the used items were simply sent down to the reprocessing unit to be cleaned, disinfected, sterilized, and reconfigured to be used again. During my time on the inpatient units, both Baluster Hospice and Hopewell Hospital did reprocess some materials. At Baluster, there was no official reprocessing department since the size of the facility was quite small. Instead, cleaning reusable materials was part of the job description of the supply clerk, Melissa. A few times each week, Melissa put on a pair of medical gloves, grabbed a container of disposable sterilizing cloths, and wiped down the durable medical equipment sent down from the patient areas for cleaning. The items included baby monitors, lock boxes used to hold patients' personal belongings, hearing amplifiers, bed alarms that alerted nursing staff when a patient with a high risk of falling was trying to get out of bed, and personal alarms that alerted staff when Alzheimer patients were wandering. It was a fairly small operation, and without any sterilizing equipment, Baluster lacked the ability to clean anything that required more than a wipe-down to be safe for reuse.

In contrast, Hopewell's reprocessing department was running 24/7, cleaning, sterilizing, and repackaging items to send back to the floors. The main items that were sent from the conventional cancer and palliative care units to be cleaned included metal IV poles, IV pumps, and some monitoring devices. In the past, the number of materials being reused on the units had been much greater. Margaret was a staff member at Hopewell's reprocessing department who could remember the time before so many supplies shifted to disposable products. "Then people started questioning the ability to clean these things well enough. So then in 2000 I think, the FDA [U.S. Food and Drug Administration] came out with their big edict on the reuse of single-use devices, so you couldn't reuse anything that said it was single use. So that's when we really started throwing things out, in the year 2000."

"So before then, even if it was single-use, you would clean it?" I asked.

"Yes, we would," she replied. "We'd clean it and sterilize it, and before then things were *made* to be cleaned and reused. I'm not saying all of it was a good idea, but even things like catheters that go into people's bladders, those were all reusable. Surgical gloves were all reusable, and I don't know that we need to go back to all that but it has been just a total reversal."

Besides changes in cleaning regulations, Margaret suggested that increased complexity of surgery and other medical services had led Hopewell away from reusable materials. "As the procedures we perform become more complex, the instrumentation we use also becomes more complex. Everything used to be pretty much metal, and now everything has flexes, and gears, and telescopes that you just can't take apart [to clean]. The challenge to clean these things and the way you sterilize is just totally different than the way it used to be. As technology changes we change how we do things, but I think we do have responsibility to think about what we're adding to the environment. I think as health care professionals it is part of our responsibility."

After a moment, Margaret added how political economy has also played a role in the switch from reusable to disposable supplies. "We used to use glass suction bottles. With what it costs now for somebody to wash those, you can't afford to pay someone what it would cost."

"Because the plastic ones are a dime a dozen and it would cost so much more to hire a person? Would it take a whole person to wash those?" I asked.

"No, because you can do so many per hour. But it's still cheaper to buy the plastic than to pay somebody, and it's not a job that anybody wants to do," Margaret replied. "I mean, it's the same type of thing as washing out bed pans, which is why bed pans are thrown away. And then you look at the bed pans. To wash and sterilize them that takes soap, that takes water, and it uses part of our environment too. So there are a lot of trade-offs."

Margaret was very conscious of the environmental impacts of reprocessing, and she suggested several things that Hopewell was doing to try to decrease that impact. Her list included the hospital's choice of reusable cloth drapes and gowns over disposable alternatives in the OR, as well as surgical instruments that were reprocessed and sterilized using reusable pans and cloth rather than disposable options. "There is still an environmental impact," she realizes, "because of the detergents and the water and everything but I tend to think it's better than throwing all of the paper and plastic into the environment."

Another environmentally friendly practice employed by the reprocessing department was a "steam under pressure" sterilization process that cleaned the hospital's reusable items. Margaret said it is "very cheap, and it doesn't really release anything into the environment, but we do use a tremendous amount of water to pull the vacuum for these sterilizers. So, I'm looking into either getting some water-saving devices or upgrading our sterilizers. But it is really hard for me to do because water is so cheap."

Margaret spoke a lot about trade-offs. She had concerns about the environmental trade-offs of the materials and processes used in her area of the hospital. She also raised the question of trade-offs for workers with her comments on the fact that no one wants a job washing out bedpans, and how workers in the reprocessing unit often encounter chunks of flesh and bloody instruments in their work.

"Have you ever had workers who are fearful of these things?"

"We ask them before they come to work here if that's going to bother them," she said, "because they do handle bloody, dirty things and if they say, 'oh my god, I can't do that,' well, probably they're not going to work here. But we do use as much automation as possible . . . we try to automate as much as we can, adding to the cost but to protect our staff plus to do a really good job of cleaning. So there's another trade-off to protect the staff." Margaret further explained that in the mid-1990s new standards on blood-borne pathogens stated that it is "the employer's responsibility to provide and make sure the employee is protected. So that's when I come down and say, hey, if you get hurt, I get in trouble. So we provide them with hair covers, plastic face shields, they have the option to wear a mask if they want . . . they wear fluid-resistant gowns. They wear nice big heavy gloves, and shoe covers. So they are protected as much as we can protect them."

This comment prompted me to ask Margaret about hazardous exposures faced by the reprocessing staff, and whether she thought part of the problem could be that the clinicians on the units did not consider the potential exposure they are creating for the next person down the line.

"Oh, man," she replied, "that is something my people face every single day. It is nothing to pick up a set that has sharp hooks in it, or sharp scissors, or someone has left a knife blade or a needle, and the scary part is when [our staff] is injured. And people do forget about that, they really do." The reprocessing staff were separated from clinicians both physically and in terms of job tasks, and they represented another group of invisible workers without whom the system would fail. Unlike supply manufacturing workers in other countries around the world, the reprocessing staff members were present in the hospital. But because of the divides between different groups of hospital staff, they remained invisibly in motion.

Cumulative Costs of Care

As previously described, the unintended consequences of medical supply chains include environmental degradation and occupational and public health risks faced by workers and communities across the life cycle of a product. These consequences accumulate according to the volume of supplies used within a health care setting, and in the case of my research sites, it is clear that the hospital used a larger volume of supplies than the hospice facility. According to institutional purchasing records, the conventional and palliative care units at Hopewell Hospital averaged thirty-three pairs of medical gloves per patient per day. In comparison, each patient at Baluster Hospice required an average of one pair of gloves per day.[13] This does not mean that clinicians at Baluster were actually walking around with a pair of gloves tucked in their pocket to reuse for each patient. The number is somewhat exaggerated because Baluster's purchasing records included all activities performed by Baluster, including at-home visits, which typically occurred only once or twice per week per patient. So the actual number of gloves used each day for each patient, particularly within the acute care inpatient unit, is higher than the analysis suggests.

However, in my observations I did notice a large difference in glove use at Baluster compared to Hopewell. While the nurses on the conventional cancer and palliative care units used gloves almost every time they were in patient rooms, nurses on the hospice unit tended to use gloves only when bathing patients, administering medications, changing wound dressings, or otherwise actually coming into contact with bodily fluids. Several factors may help explain the lower use of gloves at Baluster. The nurses at Baluster tended to be older and more experienced than the nursing staff at Hopewell. This, coupled with the idea that "there is no emergency in hospice," may help explain why nurses tended to enter a room without gloves to first determine what the patient needed, then put gloves on once they decided they were necessary. In addition, nursing staff at Baluster may have ascribed to the perception that hospice was meant to be more like home than hospital and chose to speak with patients without wearing gloves rather than approaching patients in the institutionalized gear.

These differences in institutional culture and clinical practices appeared to play a major role in determining the volume of medical supplies used in each setting. As one oncologist had suggested, conventional care uses more interventions than palliative care or hospice, and more interventions require, "more of everything else." If we wish to move closer to a sustainable health care system, we need to move beyond facilities and operational management to consider how these cultural factors drive the demand for health care.

4

Pharmaceuticals

Monica, a nurse who divides her time between Hopewell's conventional cancer and palliative care units, swipes her ID badge past the automatic door sensor. She pushes the door open, and we step inside the unit pharmacy room. It is ten o'clock in the morning, time to deliver "daily meds" to patients, and the room is a flutter of activity with the unit's pharmacist and two other nurses already preparing medications. Monica grabs a handheld "mobile meds" device that looks like a clunky, oversized cellphone, uses it to scan her badge again, and logs into the computerized medication system. The names of her four assigned patients pop up on the screen. She selects one name, grabs a disposable one ounce plastic pill cup, and begins maneuvering around the room, pulling together all the medications prescribed to this individual.

This patient is on the hematology service and has orders for a number of pills, each individually wrapped. Although the plastic packaging becomes waste, dispensing pills one at a time actually helps reduce the waste of pharmaceuticals themselves. If a physician changes a patient's medication order, the individually wrapped pills can be sent back down to the central pharmacy and dispensed to another patient, whereas if each patient were given a bottle of pills, the entire content of the bottle would need to be trashed. Monica quickly fills up the plastic pill cup with a multivitamin, a capsule of ranitidine (to reduce stomach acid), another of prednisone (a steroid immunosuppressant), one acyclovir pill (an antiviral), a co-trimoxazole pill (an antibiotic), and three gabapentin pills (to relieve neuropathic pain, a side effect from the patient's chemotherapy treatments). Once Monica has all the patient's pills, she slides open a small drawer and grabs a syringe of posaconazole sulfate, an antifungal drug for immunocompromised patients. We step into the hallway just in time to make way for another nurse who is swiping her badge to enter the pharmacy, and we head down the hall to the patient's room.

The patient is quite happy to see Monica. His breakfast has just arrived, and he knows that he is supposed to take one of his medications with food. Monica scans the patient's wristband into the mobile meds device—a safety check to make sure that the drugs she has checked out of the pharmacy are the correct ones for this patient. If the system detects an error, the device will set off an alarm to alert the nurse that something is wrong. All is well, so Monica places the pill cup on the patient's breakfast tray, pops the top off the syringe, and hands it to the man to squirt into his mouth. "I don't know what all these pills are," the patient says while looking into the cup, "but I'll take your word that I should have them." At this prompt, Monica points to each pill in the cup and tells the patient its name and purpose. He nods, empties the entire contents into his mouth, and swallows the pills down with a gulp of coffee.

Pharmaceutical Reliance

Over the past decade, global consumption of pharmaceuticals has steadily risen as reliance on medications continues to increase in high-income countries and access to health care continues to improve in low- and middle-income countries.[1] In 2020, estimates suggested that the total number of pharmaceutical prescriptions globally would surpass 4.5 trillion doses, representing over 100,000 tons of human medications. In the United States, nearly half of all Americans have used at least one prescription drug in the past thirty days. That number increases as we age: 85 percent of people aged sixty years or older use at least one prescription drug, not to mention the additional daily use of over-the-counter medications. In total, pharmaceuticals represent between 10 and 15 percent of total health care expenditures in the United States.

Pharmaceuticals are certainly among the most used resources in medical practice, and my research sites were no exception (see table C.2 in

Appendix C). According to institutional records, Hopewell's conventional cancer unit dispensed a total of 577 different drugs over the course of one year, with an average of about five and a half prescriptions per patient in addition to any chemotherapy. On the palliative care unit 270 pharmaceuticals were used in total, and on average each patient had about two medications prescribed. At Baluster Hospice, a total of 428 different drugs were dispensed, and patients were prescribed two medications on average. Similar to the volume of medical supplies used within each unit, the differences in the average number of pharmaceuticals prescribed per patient reflect the different philosophies and goals of care at each site. Many of the medications used in conventional care were antibiotics, antiviral treatments, and antifungal drugs used prophylactically to minimize infection risk among patients whose immune systems had been compromised by chemotherapy, along with drugs to relieve pain and other symptoms, including nausea and constipation. The medications prescribed to palliative care and hospice patients were typically given to control pain.

Within each of these health care settings, pharmaceuticals were considered a primary tool to help support goals of care. The same is true across medical care, in general, and pharmaceuticals have offered many benefits to improve lives for centuries. As a society, we rely on various medications to decrease pain, relieve suffering, manage symptoms, increase our quality of life, and prolong the length of our lives. Yet, as with most things, these benefits come at a cost. Just as in the case of medical supplies, at the basic level, pharmaceuticals are global supply chains representing raw materials that are harvested from the environment, processed and manufactured, transported and delivered for use, and ultimately disposed of as waste. Every pill, puff, or drip of medication represents connections to people and places these materials have encountered along the way to being used in health care treatment—but, as with medical supplies, these connections are lost and the costs to the environment, workers, and public health tend to remain hidden.

The Story of Taxol and the Upstream Costs of Pharmaceuticals

At the most upstream point in the life cycle of a pharmaceutical, environmental concerns tend to focus on ecosystem damage and loss of biodiversity that results from bioprospecting—the search for new drugs from nature.[2] The case of Taxol, a chemotherapy agent effective against breast, ovarian, and lung cancer, offers insight into perhaps the best-known story of upstream environmental impacts of a drug, while also offering an example of environmental advocacy as it relates to health care.[3] The story begins in the 1960s, with the National Cancer Institute and the U.S. Department of Agriculture working together to identify plants whose natural chemical compounds showed anticancer activity. Among the thousands of plants that were tested, the Pacific yew tree (*Taxus brevifolia*) was one of the few species whose chemical makeup showed promising cytotoxic properties. In 1967, the active ingredient derived from the tree's bark was identified through laboratory research and officially named taxol—a combination of the tree's scientific name and reference to the hydroxyl groups in the compound's chemical structure. In 1984, the drug entered clinical trials but experienced several delays resulting from a shortage of the Pacific yew bark required to produce the drug.

The tree is an extremely slow-growing evergreen species that reaches its full height of about 45 feet and average width of just under 2 feet in roughly a hundred years. Producing the amount of taxol needed to treat just one patient required the bark from as many as six trees, and stripping the bark killed the trees. At the time, the species was primarily described as a "trash" tree by the logging industry that held control of the land where the species grew in its native Pacific Northwest old-growth forests. However, as more people learned that taxol production was resulting in the death of up to 250,000 Pacific yew trees each year, calls for protecting the species began to increase.

Two key arguments were at stake. First, the sustainability of the tree species was needed to ensure the continued production and availability of taxol to treat cancer patients. Second, the environmental community saw that protection of the Pacific yew offered a way to conserve old-growth forests, which were also home to the endangered spotted owl. At the time, industrial harvesting of Douglas fir, spruce, and cedar trees for timber and papermaking was clear-cutting the old-growth forests, thereby destroying the natural habitat of both the tree and the owl. As environmental groups pushed for conservation measures, an ethical debate arose that pitted the lives of trees and owls against the lives of cancer patients, reaching its climax with a 1991 *New York Times* headline that read, "Save a Life, Kill a

Tree?" While an effort to list the Pacific yew as an endangered species failed, the species did eventually win federal protection through the Pacific Yew Act of 1992, which required sustainable management of the federal lands where the tree grew in order to ensure a sufficient supply of the tree for taxol production.

As the demand for taxol continued to increase, new production practices arose that decreased the need to harvest yew bark from old-growth forests. Various methods of obtaining the drug included isolating the active cytotoxic chemical from other parts of the tree, such as needles, cultivating the tree for taxol harvest, developing a semisynthetic process to manufacture the active ingredient using material from European and Himalayan yew species, and eventually synthesizing the drug directly in the laboratory. In time, these new developments in taxol production took the pressure off the Pacific yew. However, as the pharmaceutical industry shifted taxol production methods, the overharvesting of *Taxus contorta*, another yew species found in Afghanistan, India, and Nepal, led the International Union for Conservation of Nature to list this tree as an endangered species. This shift highlights concerns about pharmaceutical companies harvesting plants from areas where environmental regulations and enforcement are lacking. All told, the story of taxol brings to light many of the concerns related to upstream impacts of pharmaceutical production.

Drug Manufacturing, Emissions, and Environmental Health

Further downstream, the energy used in drug manufacturing poses climate-related risks to both public health and the environment. Surprisingly, the carbon emission intensity of the global pharmaceutical industry is 55 percent higher than the global automotive industry, but a lack of transparency and accountability has kept these facts hidden.[4] Pharmaceuticals are complex compounds that require energy-intensive, controlled environments with ventilation for "clean rooms" and specific temperature ranges to ensure chemical safety and longevity. These high energy demands mean that the pharmaceutical industry in the United States alone spends USD 1 billion on energy costs each year. In response to these high energy demands, the U.S. Environmental Protection Agency sponsors an industry-specific

Energy Star program to help implement energy-efficiency practices in U.S. drug companies, but these sustainability measures have not been implemented throughout the industry. Of the fifteen largest drug companies, which collectively represent 60 percent of the global pharmaceutical sector, only one-third are reportedly on track to meet 2025 emissions reduction targets previously set by the now defunct Paris Climate Agreement.

A large portion of the pharmaceuticals consumed in the United States originate in countries where environmental regulations and reporting are limited. Eighty percent of the active ingredients used by American drug companies to manufacture medications originate in India and China.[5] In India, a small number of companies within the pharmaceutical industry have stated interest in adopting sustainable practices, including life-cycle analysis and green chemistry approaches, but environmental regulations in this sector are in their infancy, and energy use and greenhouse gas emissions are currently not reported. In China, the growth of the pharmaceutical industry between 2000 and 2016 equated to a doubling of greenhouse gas emissions. Over the same period, the use of standard coal as an energy source in China's drug manufacturing increased from 6.94 million tons in 2000 to 16.37 million tons in 2016, linking the production of medications meant to support health to the extremely detrimental impacts of air pollution on respiratory and cardiovascular health.

Several air pollutants are released from drug manufacturing facilities that burn coal for energy, and very few have any form of abatement technology.[6] Particulate matter in ambient air pollution, especially fine particles with a diameter of 2.5 microns ($PM_{2.5}$) or smaller, is estimated by the World Health Organization to cause 4.2 million premature deaths globally, and more than 1 million excess deaths in China alone, from cardiovascular disease, respiratory illness, and cancer each year. Nitrous oxides contribute to the creation of photochemical smog, the telltale sign of air pollution, and can aggravate respiratory illnesses, including asthma. Sulfur dioxide pollution harms both human health by causing respiratory distress, and ecosystem health through the creation of plant-damaging acid rain. Meanwhile, many volatile organic compounds released through the use of solvents in pharmaceutical manufacturing are known carcinogens.

Again, we see how the process of creating the products that may improve health at the individual level harms health at community and even global levels. Next, we turn to see the impacts that drugs can have on health care workers at the patient's bedside.

Occupational Hazards of Chemotherapy

Nancy, an RN on Hopewell's conventional cancer unit, stops at the cabinet just outside a patient's room and puts on a pair of nitrile gloves and a thick blue paper "chemo gown." The gown is meant to protect her from any chemotherapy chemicals that might leak from the IV bag she is administering to a patient with breast cancer. I follow Nancy through the patient's doorway, and we are greeted by the patient—a bald-headed woman who is reclining in the hospital bed—and her visitor. As Nancy begins to hang the bag with the bright orange "hazardous substance" sticker on the IV pole, the visitor introduces herself to me and asks who I am.

"Oh," the patient chimes in, "she's doing an environmental study of all the stuff they use here."

The visitor replies, "That's especially good on this unit since they're pumping poisons into people."

Chemotherapy agents are a mainstay of curative cancer treatments. Most often administered intravenously, but sometimes also available in pill form, the drugs in this class are designed to interrupt cell processes, thereby killing rapidly dividing cells. Though new technologies increasingly provide targeted treatments that specifically attack cancer cells, most chemotherapy agents indiscriminately halt the cellular processes of *any* rapidly dividing cells. This explains the various side effects that patients receiving chemotherapy experience, and gives people the telltale signs of cancer treatment.[7] Hair follicles are among the body parts that contain rapidly dividing cells, and when chemotherapy knocks out these cells a person will lose their hair. Additionally, cancer patients often tire easily and may become anemic since chemotherapy also disrupts rapidly dividing bone marrow cells, which produce the red blood cells that deliver oxygen throughout the body. Further side effects of chemotherapy that result from the disruption of bone marrow include a compromised immune system and increased risk of infection that results from a lower white blood cell count. The cells that line our digestive tracts are also rapid dividers, and their short-term death from chemotherapy can lead to mouth sores, nausea, vomiting, diarrhea, and constipation. Likewise, the cells of the reproductive tract are also fast dividing, which is why chemotherapy treatments can leave cancer patients infertile. In addition, some chemotherapies can cause nervous system damage, including peripheral neuropathy experienced as numbness, tingling, and pain (especially in the extremities); damage to the heart, liver, and kidneys;

and even secondary cancers. For patients, the benefits of chemotherapy in extending life outweigh these side effects, but the potential health outcomes of exposures to many antineoplastic drugs classify them as hazardous substances.[8]

Nurses on Hopewell's conventional cancer unit administered a combined total of 1,990 doses of fifteen different chemotherapy drugs to their patients during one year of my observations, equating to about one dose per nurse per day. For comparison, nurses on the palliative care unit administered zero doses of these drugs, while on the hospice unit a total of twelve doses of five different chemotherapy drugs were dispensed in pill form, or about one dose per nurse every three months. Although patients must forgo curative treatments to enter hospice, the rules do allow patients to receive palliative oral chemotherapy to help manage symptoms and improve quality of life if the patient's hospice physician determines that such treatment would likely be effective. While coming in contact with oral chemotherapies does pose health risks, especially among at-home caregivers who handle these pills without appropriate personal protective equipment, the major concerns about the effects of antineoplastic drug exposures stem from liquid medications delivered intravenously.[9] Since the chemotherapy administered on the hospice unit was given in pill form, the potential impacts on the health of pharmacists and nurses appear to be lower than those associated with the liquid drugs delivered on the conventional cancer unit.

Chemotherapy drugs that are delivered intravenously pose several concerns to the estimated 8 million health care workers who come in regular contact with these chemicals, particularly pharmacists, nurses, and technicians.[10] Many antineoplastic chemicals have genotoxic properties, which can damage DNA and cause genetic mutations, and as a result many of these drugs are known carcinogens. The health effects of occupational exposures to chemotherapy agents among nurses and pharmacists include increased risk of chromosomal damage, decreased fertility, increased miscarriage, increased likelihood of giving birth to children with low birth weight and congenital abnormalities, and, ironically, increased risk of developing leukemia and other cancers.

I asked several nurses on the unit whether they had concerns about their exposure to chemotherapy.

"Oh yeah," Jackie, one of the nurses who splits her time between the inpatient unit and the outpatient oncology clinic, responded. "Especially since I started working in the [outpatient] clinic. I hang bags of chemo all day that say 'hazardous.' And I wear certain gloves and certain gowns around them. And I've spoken with my husband about it and we don't want that to be my permanent job. A couple years, yeah, but I don't want to hang chemo full time for the rest of my life." Jackie told me that her concern is even greater after a friend who works in the pharmacy tested her personal blood levels for trace amounts of chemotherapy chemicals and found them to be elevated. "I've never had a [chemo] spill," she told me, "but you don't need to have a spill to be exposed to it."

To protect herself beyond the standard procedure of wearing the blue paper chemo gown and chemo-rated nitrile gloves, Jackie also takes a few other precautions. "I wear my glasses at work," she said. "I don't wear glasses in my normal life . . . and I leave my shoes at work, because of chemo but also infection. I walk barefoot in my house all the time. I have a dog and I don't want to expose him to all that stuff either."

Carolyn, another nurse on the unit, told me that the gowning and gloving protocol was new on the unit and had been implemented only in the past year or so. "There were nurses who would come from other hospitals," she said, "and then come to work on [our unit] and they would just be kind of surprised that we didn't do more for protective gear. Some hospitals I think that don't give chemo so frequently take it overboard and they wear 3-inch thick gloves, and the whole nine yards. And I think it's important to protect yourself, and I think when you encounter it on a daily basis you kind of normalize it and take it for granted that you're giving hazardous drugs to people."

In addition, Carolyn highlighted the tension between patient care and nurses' health. "I also think that the way the drug is administered can affect the way the patient feels about what you're putting into their body," she said. "If you walk into the room with goggles and gown and mask, and they're like okay this is what you're putting in me and you can't even come into contact with it?" When I asked her what she thought about this tension, Carolyn responded, "I don't know. I mean, some nurses on the floor, if they're trying to get pregnant or are pregnant, they have the personal option to choose not to handle the drug. And that's totally acceptable on our floor, but you couldn't work on our floor [long-term] and not give chemo. I mean, all our nurses are chemo certified and it's expected that you give it."

Liz, another RN on the conventional unit, told me, "Just the other day, maybe I shouldn't say this, but, just the other day I had my first chemo spill. . . . I was in a hurry, so you know you're supposed to spike a bag just right so you don't puncture through it. Well, the tubing wasn't prepared just right, and I just spiked it holding it up and it sprayed all over."

"Wow, so what did you do?" I asked.

"I had the [chemo] gown on, so I just brought the IV bag and everything into myself and wrapped it all up in the gown, and I waited a minute until another nurse could help me get my gloves off and get the gown off because they were covered in chemo." Both Liz and the coworker who helped her were likely exposed to the chemical agent—this time very visibly.

One of the pharmacists who works on the unit told me, "I think the nurses to some extent understand the risk, but most of them consider putting on all of the protective gear to be a nuisance in their workflow, and so I think they probably don't utilize it as much as they even should now. I think they'll generally be opposed to increasing protective guidelines." But Tracy, a nurse on the conventional cancer unit, spoke to the contrary by saying that the current gowns were not really effective. She told me that they are supposed to prevent exposures to chemo, but in effect they just protect your clothing, "not your face, not the patient," and once the chemical does spray it gets on the patient's bed linens, the floor, and other items in the vicinity of a spill, which then expose anyone else who comes in contact with these items, including the housekeeping staff.

Chemotherapy spill events are surprisingly common. One survey asked about 1,800 nurses from across the country about their safety practices and any exposures they had experienced when administering antineoplastic drugs.[11] Nearly 14 percent reported that they had experienced an adverse event within the past week alone. The risks posed by a spill can be minimized by following current guidelines and safety protocols, including wearing disposable gowns coated with polyethylene plastic along with two pairs of chemotherapy-rated gloves—one under the gown cuff and one over the cuff. However, 42 percent of nurses surveyed said they did not always wear recommended gowns, 80 percent did not wear the recommended two pairs of gloves, and 15 percent did not always wear even one single pair of gloves.

While there appears to be some conflict between occupational safety recommendations and nurse practices, pharmacists at Hopewell were very aware of the risks that chemotherapy posed to health care workers and were pushing for a few changes in the chemotherapy delivery protocol that would limit the nurses' exposures.

As one pharmacist explained, "One of the most hazardous steps is actually the nurses attaching the tubing to the chemotherapy bag because you're essentially poking a sharp object into the bag. . . . So we've pushed for having that done in the hood in the central pharmacy where it's a controlled setting and any spill would be contained and the technician always has protective gear on. So it's a little bit safer situation for everyone involved."

Pre-priming the chemotherapy bag within the central pharmacy is one option for minimizing nurses' exposure to these hazardous substances, while identifying the barriers that prevent nurses from wearing recommended personal protective gear is another. From my time on the conventional unit, it is clear that these barriers include a complex overlap of cultural norms and time pressures—themes that are present further downstream in the pharmaceutical life cycle as well.

The Dilemma Posed by Narcotics

I shadowed along behind Marta, a nurse at Hopewell who was charged with caring for four oncology patients on the conventional cancer unit. We entered the unit pharmacy where she scanned her ID badge into a mobile meds device and signed herself into the locked, computerized medication system that dispensed controlled pharmaceuticals. Marta clicked through the computer screens to gain access to a dose of hydromorphone, an analgesic, for a young man whose colon cancer disease progression was causing him a lot of pain. After a moment, a locked drawer connected to the system automatically popped open, and Marta removed one syringe of the drug. In order to close the drawer, the system required her to count the number of syringes still remaining and enter this count into the computerized system. When the number she entered matched the expected number in the computer, she was able to close the drawer and continue on her way.

But Marta was not yet able to leave the unit pharmacy and deliver the medication to the patient. The patient needed a 1 milliliter injection of hydromorphone to control his pain, but the drug was only available in a standard dose of 2 milliliter syringes. Since this pharmaceutical is considered a Schedule II controlled substance (it has a medically acceptable use but a high potential for abuse that may lead to severe dependence), its use is tightly restricted. At Hopewell, the policy for handling such drugs required the nurse to "waste" whatever amount of the dose was not currently needed by the patient before leaving the unit pharmacy, and a second nurse was required to verify the wasting process to decrease the chance of drug diversion. At the time, another nurse was preparing meds in the unit pharmacy, so Marta asked her to verify. This is such a common practice that the nurse readily agreed, and in a minute the two were standing over a large black plastic pharmaceutical waste bin as Marta emptied 1 milliliter of the syringe's contents into the container. The second nurse verified the correct dose in the system and initialed that she witnessed the wasting process. Now Marta was ready to be on her way.

These controlled substance regulations are in place to prevent people from stealing the extra drug for personal use or to sell on the black market. As a result of these regulations, several of my field notes read as follows: "Nurse prepared morphine injection for patient, needed dose was 2 milligrams but only 4 milligram syringes were available. Wasted half the syringe!" A single occurrence of this wasting practice does not seem to be of much concern, but consider the volume of medications wasted over a year and the cumulative impacts of this practice begin to emerge. Such waste reflects several potential environmental and public health impacts since the drugs represent numerous upstream resources used in their development, manufacturing, and delivery, as well as a number of downstream impacts from their disposal through incineration.

Environmental Impacts of Hazardous Waste Incineration

Once a day at Hopewell, an Environmental Services staff member came through the unit to place a locked lid on the black bin and remove it from the unit pharmacy to the basement. The bin contained all the drug waste generated by the conventional cancer and palliative care inpatient units at Hopewell, including anything from chemo IV bags to liquid narcotics to pills. Once the bin was removed to the basement, it was stacked with all the other black pharmaceutical bins from around the hospital and kept under lock and key in a storage unit behind what resembles a metal garage door that rolls down from the ceiling. Once the storage unit was full, the hospital's hazardous waste hauler would pick up the whole load of bins—for a total of just over 74 tons a year for the entire facility, including all inpatient units, outpatient clinics, and the central pharmacy. Just as in the case of the other forms of waste, no one seemed to know where the waste was destined. One of the Environmental Services administrators gave me the number to call to find out.

From the hazardous waste hauler, I learned that the black bin waste was transported to a statewide holding facility, where it joined similar waste from other health care institutions—including Baluster Hospice, though at a fraction of the volume[12]—and again waited until enough bins were gathered together to fill a semitruck. Once a full truckload was compiled, the bins were transported to a certified hazardous waste incinerator in Port Arthur, Texas, for destruction—plastic bins and all.

Similar to the incineration of infectious waste described in chapter 2, hazardous waste incineration poses several environmental and public health threats.[13] Emissions from hazardous waste incinerators include the heavy metals cadmium and chromium, as well as highly toxic dioxins and furans, and organics such as polychlorinated biphenyl, benzene, and toluene. Epidemiological research shows that people living close to incinerators experience higher rates of all cancers than the general population, with the clearest evidence linking hazardous waste incinerator emissions to non-Hodgkin lymphoma, soft tissue sarcomas, and lung cancer. Since hazardous waste incinerators and landfills are disproportionately sited in minority communities, the public health impacts of these facilities are an issue of environmental justice and racism.[14]

The Port Arthur incineration facility is no exception. The city itself is home to just over 53,000 residents, 40 percent of whom identify as Black or African American, and 23 percent of households live below the poverty line.[15] The community where the hazardous waste incinerator is located, along with several petrochemical plants (including the largest oil refinery in the United States), is reportedly 80 percent Black or African American and Latinx. Air pollution is among the greatest environmental health threats to residents in the area, compounded by a high poverty rate and limited access to fresh, healthy foods; safe, walkable neighborhoods; health care; and other social determinants of health. The hazardous waste incinerator that destroys Hopewell's waste is certainly not the only challenge facing the people who live in Port Arthur, but the fact that our health care systems rely on incinerators in low-income communities to destroy our waste adds to the irony of how medical care at the individual level affects health at the community level.

And yet Port Arthur is not where the story of hazardous medical waste ends. The toxic ash from the incinerator is transported from Texas to Emelle, Alabama, the home of the largest hazardous waste landfill in the country, referred to as the Cadillac of Dumps.[16] From the time the pharmaceutical waste left Hopewell or Baluster's loading dock to the time it was landfilled as toxic ash, the materials had traveled over 1,600 miles and passed through many hands and places before reaching its final resting place. Just as in the case of Port Arthur, the hazardous waste landfill in Emelle represents another environmental justice challenge. In this rural community of just over fifty people, 94 percent of residents are Black or African American, and 62 percent of households live below the poverty line. When the landfill was created in 1978, there were no Black representatives on the county industrial development board, on the county commission, or in the state legislature, providing evidence of environmental racism in the siting of this facility. While no clear alternatives exist for the safe disposal of hazardous pharmaceutical waste that would have lower environmental and public health risks than incineration and subsequent landfilling of toxic ash, there are means by which the health care industry could limit the amount of waste generated.

Competing Interests in Pharmaceutical Safety

I asked a pharmacy administrator at Hopewell to explain why so many drugs were wasted into the black bins. He told me, "There is a legal requirement to document controlled substances to account for all elements of the dose. So if you only need seven and you take ten, there has to be a witness to the wasting of the three milligrams. And those three milligrams that you are wasting have to become nonrecoverable." Hence the black plastic drug bins in the unit pharmacy where all the wasted drugs mingle together. When I asked if it was possible to get data on the total amount of drugs being wasted each year, he replied, "I probably would not do a detailed data analysis of that. I would say it's probably less than 10 or 15 percent. The other thing I would say is that narcotics for the most part are relatively inexpensive; I don't think they represent a huge environmental hazard. I don't know how interesting that really is. People seem to be really worried about narcotics. But I think the bigger worry is the ability for it to be diverted, but that's why we have the controls and verification of its waste."

Concerns over drug diversion and the illicit use of narcotics are reasonable, especially considering the number of lives affected by the opioid epidemic.[17] But at Hopewell, if 10 or 15 percent of all the narcotics dispensed are being wasted, is such inefficiency in the system worth looking at more closely? I asked this administrator whether it is possible for the manufacturer to provide the drugs in different doses that would eliminate this waste.

"The manufacturer will generally make the most common form," he suggested, "but the physicians try to individualize the doses as much as possible. For the vast majority of times we administer something it is the standard dose, but narcotics are kind of unique. For example, morphine, I think it comes in 10's, 8's, 4's, and 2's [milliliters]. Something like that. So if you need seven [milliliters], you can't really add those up."

Innovations in the volume of liquid medications available are one way to help overcome the challenge of minimizing narcotics diversion, though nursing staff at Baluster Hospice's inpatient unit had developed a different strategy.

"Sometimes if a patient has 1 milligram ordered and it comes in a 2 milligram [syringe]," Samantha, an RN on the hospice inpatient unit tells me, "I sometimes will save that milligram to use later for that patient on my shift. But before I leave my shift, if I haven't used it, I will waste that milligram with another person [to verify] because I don't want it hanging around. If I can save it and then use it that shift, it's not going to be wasted. But otherwise it is still squirted down the sink."

Two new insights come to light here. First, institutional policy allowed nurses at Baluster to save doses of controlled medications for use later on that same shift for the same patient, thereby minimizing waste. Second, when a liquid drug, such as morphine, was wasted on the inpatient hospice unit, rather than being dispensed into the black pharmaceutical waste bin, it was instead flushed down the sink where it enters the wastewater system—a practice that one hospice physician suggested led to "happy fish."

When I asked one of the hospice pharmacists whether such disposal practices should be of concern, he said he was less concerned about controlled substances entering the wastewater system compared to more toxic substances. Narcotics "tend to be substances that break down fairly readily," he told me, "and they're actually closely related to a lot of natural products. I mean the original opiates were from opium, which is from poppies. So it's a naturally derived structure, and most plants and animals have some sort of [ability to metabolize these substances]. So from that standpoint I have less concern unless that level were to get really high . . . but other medications I think yes, there are potential risks."

A deep look into the literature on pharmaceutical pollution and aquatic toxicity shows that the constant and ubiquitous introduction of drug waste into lakes, rivers, and streams means that medications have "pseudopersistence" in the environment; even though they may rapidly break down, they are continuously reintroduced and so are always present. As I quickly learned, narcotics are not the only medication being flushed down the drain, and the presence of pharmaceuticals in the environment poses big threats to aquatic organisms.

At Baluster Hospice's inpatient unit, nurses flushed narcotics down the drain to comply with the Drug Enforcement Agency's rules on drug diversion. In comparison, nurses serving patients through Baluster's home hospice service flushed almost all unused drugs down the drain as a service to the family after a patient had died. Sue, an administrator on the hospice inpatient unit who had served as a nurse in the in-home setting, described it this way:

> [Baluster's] *policy* is to destroy the narcotics or controlled substances when somebody dies. The *practice* is that we offer to get rid of *everything* for them, just as a service while we're there. Pills either go down the toilet or down the sink. If it's a fentanyl patch, like a morphine patch, those we . . . put on gloves and then cut them up. . . . If it's liquid we just dispose of it down the drain or toilet. Suppositories we just flush them, too, just pop them out and dispose of them that way. IV medication we cut the cassettes and drain them down the sink or toilet. Pretty much everything just goes down the drain.

When I asked how many pharmaceuticals were actually disposed of this way, Sue replied, "There are a ton of medications being wasted" in the in-home setting.

Overprescribing and Drug Waste

The practice of disposing of unused medications down the drain is fairly common.[18] In recent surveys of drug disposal behaviors among the general public from across the country, up to 28 percent of people report flushing leftover pharmaceuticals. Sometimes medications accumulate and need disposal when a consumer purchases over-the-counter medications in large volumes that then expire before they are completely used. Other times, patients may not follow doctor's orders to complete a full course of a prescribed drug. Among hospice patients, Sue explained that by the time a person with a cancer diagnosis enrolls in the program, they have typically been through years of conventional treatment in which they have tried multiple medications, many of which end up as waste.

> There can be so much overdispensing. You would go out to a home and . . . I'd joke that I could have retired on the amount of OxyContin if I'd sold it. Because you could get bottles with 200 or 300 tablets in it and they'd take it twice and not like it or have a bad reaction to it and then be switched to another medication. And then that bottle can't be used by anybody else. And they are expensive medications so they'll hang onto it until the end and then we end up destroying it at the time of death.

Sue suggested that the majority of these medications accumulate from the time that a patient was seeking aggressive care in the conventional medical system because, in hospice, "we try to limit what we order, we approve two weeks at a time for any narcotics [to decrease waste] and the cost as well."

I observed the best example of the way that hospice manages medications during new patient intake interviews. One of the physician's very first tasks when admitting patients to Baluster's care was to examine their current medication list and determine if any of the prescriptions were no longer appropriate. This practice arose from the requirements that patients must meet in order to enroll in hospice: they must have a prognosis of less than six months of life, and they must agree that they will no longer seek curative treatment. Given these two requirements, many pharmaceuticals are no longer appropriate to provide care to the patient. Drugs like statins, which provide long-term cholesterol-lowering benefits, are usually no longer appropriate and in fact may be harmful to the patient at this point in their disease process, so the admitting hospice physician will review the medication list to discontinue any inappropriate drugs. At the same time, many patients arrive at Baluster in severe pain, and the hospice physician, pharmacist, and nurse must work together as a team to select an appropriate type and dose of medication to relieve pain.

Sue further explained how hospice tries to help reduce the amount of medication that becomes waste. "You don't need to prescribe a year's worth of medication, or even a month's worth if the patient is just trying the med for the first time or if they are a terminal patient. You know, do it week by week or every two weeks. But you also want to balance it with having to go to the pharmacy constantly."

Shannon, an organization-wide administrator at Baluster, explained a bit more about differences in practice that lead to pharmaceutical waste. When a patient gets to hospice, the philosophy is that the organization should "really be mindful of how many medications we're ordering for people because it's costly. It's costly for us, it can be costly for the patient and family, and then we end up wasting them." She paused, and then continued, "But, you know, [as a society] we're into convenience. Doctors don't want to be called for refills. Patients and families don't want to run to the pharmacy to get more . . . we've been programmed as a society that if you buy more, it will cost less or it will save me the time to have to go back and get it. So I think that plays a big part in getting more and then not using it and just throwing it away."

The Environmental Fate of Pharmaceutical Waste

Pharmaceuticals can enter the environment through several routes.[19] Some chemicals may be discharged through the drug manufacturing process. Others are disposed of down the drain by consumers. The majority of pharmaceuticals that enter the environment are excreted through urine or feces, with between 30 and 90 percent of each dose passing through the human body, depending on the specific composition of the drug. Regardless of the source, when these drugs are flushed down the drain, most pass right through municipal wastewater treatment facilities and are discharged directly into lakes and rivers. As a result, pharmaceutical pollution has been a concern for decades. In 2002, The U.S. Geological Survey published a report showing that 80 percent of the 139 streams they sampled across the country had pharmaceuticals in the water. Since then, these chemicals have been detected around the world in surface waters, groundwater, marine environments, and estuaries. Since many cities draw their drinking water from these sources, and since standard water treatment does not remove these chemicals, it may not be surprising that pharmaceuticals have also been found in trace amounts in municipal drinking water.

The presence of pharmaceuticals in the environment poses a number of potential threats to the environment and human health.[20] Drugs are designed to treat specific illnesses and can remain biologically active once released into the environment, where they can cause physiological effects, behavioral changes, and reproductive failure in wildlife. Birth control drugs are perhaps the most well-known class of pharmaceuticals that have an effect on wildlife. Reports of the feminization and reproductive failure of male fish exposed to these endocrine-disrupting chemicals have reached national headlines. Among the many medications used to treat cancer patients, anticancer drugs are of particular concern because of their toxic properties. These drugs are designed to prevent tumor cells from growing by targeting cellular processes. When aquatic species are exposed to these chemicals, a variety of ecotoxic effects can result, including cell mutations that may lead to cancer, or teratogenic effects that lead to malformation of offspring.

Another class of drugs known to have negative impacts on wildlife are antianxiety and antidepressant medications.[21] These drugs are of interest here because they are widely prescribed to cancer patients, with rates of prescription among cancer survivors nearly twice those of the general public. They also happen to be fat-soluble and tend to bioaccumulate through the food chain, as evidenced by the high concentrations of these chemicals found in wild fish. Exposure to these drugs can affect the health of species throughout the food chain. Among the species at the base of the food chain, exposure to fluoxetine (Prozac) among amphipods—small, shrimplike crustaceans—will lead these small creatures to swim toward areas of light, which could decimate their population since they tend to hide in shaded areas to avoid predation. At higher levels of the aquatic food chain, exposure to fluoxetine increases the rate of mutations, such as curved spines and a lack of pectoral fins in fish embryos, decreased growth in juvenile fish, and decreases in protective behaviors in spawning grounds among adult males. These various impacts on aquatic species suggests the potential for population crashes among exposed species, which is concerning for both ecosystems and the people who depend on them for their health and well-being.

Conflicting Guidance on Drug Disposal

With all these potential environmental and human health risks of pharmaceuticals in wastewater, why do so many drugs get flushed down the drain through Baluster's in-home care? Since home hospice nurses did not have ready access to black boxes for disposal, and transporting drugs back to the facility for disposal would be extremely problematic because of concerns over drug diversion, the practice of flushing drug waste had become the primary route of disposal.

The preferred way to dispose of leftover medication, as suggested by the U.S. Department of Justice Drug Enforcement Administration, is to return them to a take-back program where they can be destroyed through incineration, though as outlined earlier, incineration poses several risks to the environment and public health. According to the U.S. Food and Drug Administration, when take-back programs are unavailable, the recommended action is to first check to see if the medication is on the FDA's "flush list."[22] If yes, consumers are directed to dispose of the drugs directly down the drain. If no, the recommended action is to place the drugs in the household trash after mixing them with an undesirable substance, such as coffee grounds or kitty litter, in order to prevent them from being recovered for illegal diversion or accidentally poisoning children and pets. Unfortunately, this disposal method does not reduce the risk of pharmaceutical pollution since over time the drugs filter down through landfills and are eventually released into the environment through leachate.

In contrast, the U.S. Environmental Protection Agency directs consumers to use take-back programs where available, and otherwise to follow the same steps for mixing unused drugs with undesirable material and placing it in the trash—and they explicitly state that drugs should not be flushed down the drain, "unless the label or accompanying patient information specifically instructs you to do so." These conflicting directives suggest a trade-off between environmental risks of pollution and the public health risks of drug diversion that institutions and individuals must navigate.

Staff members at Baluster were well aware of the recommendation to mix drugs with undesirable substances, but I repeatedly heard nurses, physicians, pharmacy staff, and administrators say that these methods were not realistic. In particular, the hospice staff pointed out that following this guideline would require nurses to carry kitty litter around when visiting patients' homes.

I asked a number of people about the governance that leads to the different pharmaceutical disposal practices employed at Hopewell Hospital—where *all* drug waste goes in the black bins—and Baluster Hospice—where on the inpatient unit narcotics go down the drain and everything else goes in the black bins, but for in-home patients all medications go down the drain.

One of Baluster's pharmacists responded, "The [Environmental Protection Agency] is against medications going into the water system, but the [Drug Enforcement Administration] doesn't care, and they prefer that the medications are wasted [down the drain] from a security standpoint."

"So that's why the narcotics are wasted down the drain?" I asked.

"Yeah," he replied. "The DEA scares us more than the EPA. They have guns and can put you in jail. The EPA might fine you."

In comparison, Hopewell is able to easily secure the black pharmaceutical waste bins in their unit pharmacy, and is thereby capable of complying with the opposing guidelines they are given from several agencies. "I think in the hospital that's one of the more complex areas," the Baluster pharmacist replied. There are people in the pharmacy department whose jobs focus on trying to make sense of all the different guidelines "because you typically get one set of guidelines from the governmental agencies that are designed to protect employees—your OSHA [Occupational Safety and Health Administration], NIOSH [National Institute of Occupational Safety and Health] guidelines that are basic but helpful—and then you get a completely different set of guidelines from the Joint Commission and other accrediting bodies." In effect, having so many different guidelines "leads to confusion sometimes," he continued, "because you're trying to balance those regulations."

I asked Shannon for an organizational perspective on how Baluster could minimize the impacts of pharmaceutical waste. "I think it would need to be systemwide," she responded. "I think [Baluster] wants to do the right thing, but there are not good systems for meds disposal. That's really a larger health care and regulatory concern. I think we would be more than willing to participate in any efforts where we as a health care community could address that issue."

5

Patients

Dr. Johnson, an oncologist serving as the attending physician on Hopewell's conventional cancer unit, sat chit-chatting with me in the oncology conference room early one Friday. We were waiting for the rest of the oncology team—a fellow, a resident, and a physician assistant—to arrive for the group's daily meeting where the team shared updates on patients before making morning rounds. When everyone had assembled, Dr. Johnson launched into discussing plans of care for each of the five oncology patients currently admitted to the unit. One of the patients, an elderly woman whose colon cancer had not responded favorably to standard treatment, was now participating in a clinical study for a new chemotherapy drug. But things were not going well. A large hematoma had developed in the woman's abdomen, which precluded her from continuing with the study.

The team discussed this new clinical development and stated that it meant the patient's prognosis had worsened. From their earlier conversations with the patient and her family, the clinical team believed the patient's husband knew that the hematoma was "bad news." The team did not discuss whether they had suggested hospice as an option for the patient, but they did say that now the goal of care was to transfer the patient to a skilled nursing facility (a specialized nursing home) for rehabilitation, with the hope that she would then be able to continue with the chemotherapy trial. Later that morning, I followed along with Dr. Johnson as he checked in on the patient, spending about five minutes listening to her chest and abdomen with a stethoscope, gently palpating the hematoma, and asking if the patient or her husband had any questions.

The husband leaned toward the doctor and asked, "So, what's the situation here?"

Dr. Johnson responded that the patient would be discharged soon and that the team was working on getting the placement all set at the skilled nursing facility. He made no mention of prognosis or other options, even though open communication about the likely course of the patient's disease may have given the patient the choice to consider whether she would prefer to transition away from conventional life-prolonging therapy and toward hospice.

The Benefits of Hospice

Patients derive a number of benefits from having big picture conversations about end-of-life care with their health care providers. For example, patients who discuss their end-of-life preferences with their physician have lower physical distress, are more likely to receive care consistent with their wishes, and are more likely to achieve their preferred location of death than patients who do not have these conversations.[1] Understanding end-of-life options can also help patients transition to hospice early enough to derive the benefits afforded by this model of care. Patients who enroll in hospice earlier in their end-of-life care trajectory report higher satisfaction with care than those who continue with conventional care and receive life-prolonging treatments, even after it would be medically appropriate to transition to hospice. Reportedly, this increased satisfaction can be attributed to the hospice philosophy that views death and dying as natural life processes and works to preserve patient dignity and autonomy through the final stages of life.

Surprisingly, patients who enroll in hospice earlier in their treatment live for up to a month longer on average, and have higher reported quality of life compared with those who remain in conventional care for a longer period. There are several explanations for the increased longevity of hospice patients. One attributes a longer life to hospice's psychosocial care and palliative therapies. Another suggests that the social support provided by hospice's interdisciplinary team may increase a patient's desire to live, while decreasing fears that they are a burden to their family. Another suggests that the risk of mortality is reduced by the lack of high-risk aggressive conventional treatments such as late-stage bone marrow transplants or high-dose chemotherapy. Regardless of the reason, people who transition to hospice early in their end-of-life care tend to live longer than those who remain in conventional care.

Patients are not the only ones who benefit from timely transitions into hospice. Families also benefit in a number of ways. When a patient enrolls in hospice, their family gains access to a variety of bereavement services that can decrease caregiver stress.[2] Families of cancer patients who die in hospice are more likely to rate the care received by their loved one as "excellent" than those who die in conventional hospital settings, and are more likely to report that the end-of-life care provided was consistent with their loved one's wishes. Perhaps surprisingly, the social and spiritual support provided by hospice may also decrease mortality among bereaved spouses (the "widower effect").

The potential benefits of hospice care also extend to the institutional level. Hospitals benefit from hospice programs through decreased costs, decreased resource utilization, and increased quality of care and hospital performance, including shortened length of stay and decreased patient readmission rates—all of which increase profitability.[3] These measures of performance also link back to reduced impacts on the environment. As the previous chapters have shown, each of my three sites had environmental and public health impacts that extended beyond their facilities. However, my observations and institutional data show that overall, conventional cancer care uses more medical supplies and more pharmaceuticals, and it creates more waste than either palliative or hospice care. Echoing back to one of the oncologists I interviewed, "There are just more interventions with conventional care; so more interventions means more of everything else."

As a result, the longer a patient remains in conventional care when it would be medically appropriate to transition into hospice, the greater the cumulative impacts on planetary health. Nationally, many patients receive conventional cancer care right up to death, and many die while receiving care in the hospital—whether in a conventional inpatient cancer unit while receiving standard treatment, or in the emergency room or intensive care unit after an acute episode associated with their diagnosis. The poor quality of life and high economic cost associated with dying in conventional care are out of line with patient preferences and institutional expenditure goals and have been major drivers in health care reform. By expanding our view of end-of-life medicine beyond the patient level to include the ecological outcomes of care described in the previous three chapters, we begin to see the broader extent of these impacts.

Given these potential benefits of hospice as an option for providing high-quality end-of-life care, it is surprising that so many patients enter hospice programs late in their care, if at all. To enroll in hospice and receive the benefits it affords through Medicare, a physician must declare that they expect a patient has six or fewer months to live given the normal progression of their disease. While patients may receive the hospice benefit for more than six months as long as they still qualify, the median length of stay in hospice before death is just eighteen days.[4] More than half of patients enroll in hospice care less than one month before they die, and 28 percent receive hospice care for less than one week before death. According to the National Hospice and Palliative Care Organization, this short length of stay is "considered too short a period for patients to fully benefit from the person-centered care available from hospice." Overall, hospice is an underutilized service since only 52 percent of all eligible patients enroll in the program each year in the United States. This means that nearly 170,000 cancer patients receive hospice benefits for less than one week before dying, and an additional 180,000 never enroll in hospice services at all.

This relatively short stay in hospice provides evidence that, in general, patients are remaining in conventional care for longer than is medically beneficial. One recent study found that 24 percent of cancer patients received their final chemotherapy within one month of death, a period that is associated with a higher risk of receiving mechanical ventilation, CPR, and tube feeding within the final week of life, and an increased likelihood of dying in the ICU.[5] Without the support of hospice, patients are more likely to have a negative health care experience and are less likely to die in their preferred location, which up to 60 percent of cancer patients report is in their own home.

The Barriers That Prevent and Delay Transitioning into Hospice

There are several reasons why eligible patients do not transition into hospice at a time when it would be medically appropriate. Since the benefits of hospice extend to patients, their families, and the planet, identifying the barriers that prevent patients from receiving hospice care can help improve patient end-of-life experiences while also minimizing the unintended environmental consequences of cancer care.

These barriers can be complex. Although Medicare covers the cost of equipment, medication, and support services provided by hospice, not all insurance companies offer these same benefits. When a patient's health insurance provides only limited hospice coverage, they may fear leaving their family in debt if they enroll in the program, so they remain in conventional care to avoid the financial burden. At other times, patients fear that by enrolling in hospice they will no longer receive care from the health care providers they have come to know and trust, and will instead be "dumped" into a strange new system at a time when they feel most vulnerable. At still other times, patients, families, or physicians may not feel ready to "give up" the hope of finding a cure and thus refuse to entertain the idea of hospice. While I did encounter each of these barriers during my time at Hopewell, clinicians most often told me that the key to patients transitioning away from life-prolonging conventional care was "having those big picture conversations with patients earlier." The American Society of Clinical Oncology, a national organization for oncology physicians, echoes this sentiment in an official statement that suggests desisting from aggressive life-prolonging treatments when they are no longer beneficial requires cultivating open communication between physicians and their patients.

Allowing space for difficult conversations about prognosis can empower patients and families to make better-informed decisions about how they would like to spend their final months, weeks, and days. In turn, open dialogue about death and dying could also drastically reduce the unintended environmental and public health consequences of end-of-life care. But prognosis is a difficult thing to determine, and even though clinicians may privately believe they have a reasonable estimate for how long an individual patient will live, it is difficult to be certain enough to share such information with the patient.[6] I heard several clinicians in all three of my research settings support this concern, and many stated that unless a patient is actively dying, they find it hard to address the subject.

The Challenges and Opportunities of Open Communication

On a Monday, I shadowed Dr. Smith, an attending physician for the palliative care team at Hopewell. There were only three patients on the palliative care unit that day, but the team had been called in to provide consultations to patients who were receiving care on a number of other services throughout the hospital. One of these patients was a man in his late twenties suffering from testicular cancer that had metastasized to his brain. The plan was to discharge the patient back home, and the specialty team expected he would die within a few days, though it was unclear whether anyone had discussed this prognosis with the patient. The palliative care team was called in to discuss goals and options for symptom management with the patient and his family.

Dr. Smith began the conversation by introducing the members of the palliative care team and telling the patient that their job was to help patients facing "life-limiting illness" make decisions and manage symptoms. Next, Dr. Smith asked the patient and his family what their understanding was of what was happening, and how much time they had left together. The patient sat quietly in bed, so a family member spoke up to say that they were hopeful that miracles can happen and that the patient would get well. Over the course of the next fifty minutes, Dr. Smith helped the patient and family articulate their goals to be at home together, and suggested that hospice could provide a number of resources to help, although no one ever shared the estimated prognosis that the patient likely had only days to live.

This case was among the clearest examples I observed of the palliative care team being called in very late in a patient's plan of care. So late in fact that it was difficult for the team to build rapport and trust with this family. At the end of the day, I asked Dr. Smith why the team had not spoken openly with the patient and family about prognosis, even though the clinicians had discussed it several times among themselves, and particularly since the family seemed not to know this information. He paused before responding that perhaps the palliative care team is at times just as guilty of not sharing prognoses as other medical teams. Giving bad news is difficult under any circumstances, he told me, and you really only want to "drop the bomb" once, so you had better do it well, a feat that is particularly difficult when you have only met a dying patient in their eleventh hour.

Dr. Smith proceeded to tell me a story of a previous patient he had worked with who just could not understand the plan of care that the team had set forth for her. Through their conversations, the doctor eventually realized that no one had openly discussed prognosis with the patient. When he finally told her that they had created this plan of care because her cancer was killing her and that she would likely die soon, a "light bulb went off," and the patient suddenly understood that she had a new goal of care and that this plan would help her achieve that goal.

Dr. Smith suggested that in medicine, we "use a lot of euphemisms." He gave the examples, "she's very ill," "this might not get better," "things are very touch and go." He suggested that with these vague terms, "the family doesn't seem to understand that she's dying, because no one has used the 'D' word. We're saying she's really ill, she's very critically ill, she's not responding to treatment. All of those things are bad, we all know that, but it doesn't translate."

So when the palliative care team is called in for a consult, they always ask a few important questions: What have your doctors told you? Has anyone talked to you about how long this might go on? And often they find that the other teams have not gone so far as to tell the patient that they are dying.

But when the palliative care team asks what the patient and family think is really happening, the patient will often say that they think they're dying. And the family, when asked if they think the patient is dying and has only a limited time to live, will express that this has been their sense of what is happening. Other patients will express their suspicion that their doctors have been trying to tell them they are dying, but unless someone states outright that they are dying, they will continue hoping that there is still time, or that maybe the next treatment will work.

Once the patient and family have an opportunity to openly discuss that the patient is dying, the team is able to help them articulate their goals for how they wish to spend their remaining time, and how to best achieve those goals. Although not always the case, patients who have these big picture conversations are more likely to transition away from life-prolonging treatments and toward care that affords them the benefits discussed earlier. Those who do not have these conversations are more likely to continue with life-prolonging treatments hoping for a cure, bouncing in and out of the hospital when they suffer acute episodes related to disease progression or side effects of treatments. Oftentimes, they will require intensive care.

Relationship-Building in Support of Patient Care

Joanne, a social worker on the palliative care team, shared her experiences with patients and families in the ICU. The most recent research available suggests that nearly one quarter of ICU patients are admitted for cancer-related diagnoses, 47 percent of whom die in intensive care.[7] Joanne said that in many of these cases, the palliative care team has been called in for a consultation with families of patients who have been in the ICU for over a month where, when learning about palliative care options, the family says, "Oh, thank God. We don't have to keep him going like this?"

"They need permission," Joanne said. "They need somebody to let them say, 'Look I don't think we're doing the right thing here, what do you think?'"

She said that the specialty team doctors are just going in each day and giving the patient and family information about blood pressure and other vital signs, but

> that doesn't give you an opening to say, "I think we should stop . . . this is horrible, he would never want this." But if nobody gives me that out and agrees with me, or brings it up to start out with, I'm just going to go with it. I really think it's a matter of giving people the choice. . . . I think that's what the palliative care doctors do best, it's those communications and doing it the right way that takes the onus off that poor person who feels like they're all alone in the world trying to make that decision by themselves. You know, to really make it known that they're not alone and we're on their side and it really makes sense to choose either option.

As these cases suggest, the palliative care team is a key resource for helping patients and families articulate their goals. But the team's training in how to approach difficult conversations with patients may be one of the reasons why primary care and specialty teams do not discuss death and dying earlier in a patient's care.

According to Dr. Brown, a physician at Baluster Hospice, perhaps having the palliative care team as a consult service outside the regular practice of care is

> actually a bad thing, because it takes responsibility away from the people who are making the decisions earlier. I know from going to the palliative care conferences at [Hopewell] that they talk a lot about how they'll get a consult to come and talk to this [family member or other caregiver] about why they should stop the ventilator, and it's like whoa! That's not [the palliative care team's] job, that's your job! Don't you have a relationship with this person, and don't you care about communicating with them about it? It's like [the

> primary care and specialty teams] would like to just divorce themselves from the whole phenomenon and just make it go away.

"And why is that?" I asked.

"Because it's hard, it's uncomfortable," Dr. Brown replied. "So when people who embrace the palliative care model say let's do it well, others say, okay, you do it well and then I don't have to do it at all. And so it's a responsibility that people would like to relinquish because it's uncomfortable and difficult and it can be very heart wrenching."

Hopewell's palliative care service was designed as a resource for conventional care clinicians to consult when a patient needs additional expertise as they approach the end of life. But this model may inadvertently set the stage for a lack of communication between the patient and their conventional clinical team. As a result, conventional care teams called upon palliative care only when they felt there was no other life-prolonging option available for a patient. This evidence suggests that oncologists are good at telling people when their cancer is no longer curable but is treatable, but they are not as good at stopping life-prolonging treatments once they have been started.

But this is not the only palliative care model being implemented in American hospitals. In a different model, palliative care is offered alongside conventional services from the time of diagnosis. A pharmacist whose workload included spending time with both conventional and palliative care teams at Hopewell described such a system when I asked whether he thought it would be possible to implement palliative care earlier in patient care:

> I think there are some interesting practices being studied out there. . . . I was at [a conference last spring] where they had palliative care physicians seeing patients in the oncology clinic on a parallel track while they were still receiving their oncology treatment. The patient would see the oncologist for their chemotherapy and their palliative care physician for management of pain or nausea or other symptoms . . . and building that relationship to the point where when they weren't benefiting anymore from chemotherapy they would meet jointly with both groups of physicians and eventually switch over to just the palliative care track. And they found that the patients had a fairly high satisfaction with it, and the oncologists had a high satisfaction with it as well because the patients already had that relationship built so it was easier to let go of active [aggressive] treatment and go the palliative route because they

> already knew the palliative care physician and had already maybe had a positive experience with them, so I think people are looking at those sorts of things.

Dr. Robinson, a physician who shares his time between the conventional cancer and palliative care services at Hopewell, described how he was already implementing such an integrated approach in his own practice—a very local scale intervention that he suggested had helped transition many patients to palliative care at appropriate times.

"It helps to introduce hospice before you're recommending it," he said. "Maybe at the second [meeting with the patient] I say the three options—standard chemo, clinical trial, palliative care—and I will say, 'And one day I hope that we'll have the opportunity to work with hospice. I want you to know that hospice is an organization that I believe in strongly, that supports quality of life . . . and I believe in it and it's a great organization, and I hope one day we get to work with them.' Period." And throughout the course of treatment, Dr. Robinson will again reiterate his belief in working with hospice but would clarify that "we can't work with hospice right now because we're still doing chemotherapy. And just because of the way the [Medicare] rules are written, we have to wait until we're done." Sometimes he would tell the patient, "I wish we could work with hospice now because there are so many benefits to it. But because of the way the [Medicare] rules are written we can't," a fact that he believed was a real travesty.

He continued on by saying that with this parallel track approach, by the time a patient's disease is no longer responding to aggressive treatment, hospice and palliative care options have been a continuous part of the conversation. But still, when Dr. Robinson suggests that it is time for a patient to seriously consider hospice, some patients will say they are not yet ready for hospice because they are not dying. In response, he told me that he will say, "'I agree you aren't dying today, you aren't dying next week, probably not even next month . . . but you do have a serious illness and you are dying from it . . . [and because of this] you can get hospice.' I try to frame it as a right, or a privilege. 'You have a right to access hospice, and with it come all these other ancillary benefits. They pay for medications, they pay for equipment.'"

He also tells patients that hospice staff will "support your family . . . [and by signing up early] you get to develop a relationship with the nurse. And that is someone who will take care of you, quite frankly, for the rest of your life. And it will be quite nice for you to know them and trust them. So that when you are really sick and you do really need them, this won't be a stranger coming into your house. It will be someone who you've known for some time, that you have a good relationship with, that you trust, that you like."

"And that gets them to sign on?" I asked.

"I think that's what seals it," he replied. "They realize that 'yes, I don't need them today.' But if I've done my job right they realize that they will need them at some point in the future."

If he has done his job right, he has opened the conversation about death, dying, and end-of-life options. He has given a patient agency in their end-of-life care and provided the choice of whether and when it is appropriate for that person to transition into hospice, perhaps allowing that individual and their family to benefit from hospice services for a longer period of time than may be typical. In turn, the environment and public health may benefit from the transition away from materially intensive conventional care and toward interpersonally intensive hospice care in a way that offers a win-win scenario for both the patient and the planet.

6

Conclusions and Practical Implications

Early one summer morning, I shadowed Samantha, a nurse at Baluster Hospice's acute care inpatient unit, as she provided care to a middle-aged man dying of lung cancer, and his family who sat in vigil. Samantha checked in on the patient and family every thirty minutes or so to see if there was anything they needed. As the patient began showing signs of approaching death—cold hands and feet from the slowing of blood circulation, mottled skin coloration from a lack of oxygen, and eventually a shallow rasping breath—the number of family members called to his bedside grew to nearly twenty people.

About two hours after I arrived, the man died. The family was welcomed to stay with their loved one for as long as they wanted, and various staff members, including nurses, the chaplain, and a social worker, offered support over the following hours. When the family was ready, two nurses washed the man's body, and then draped one of the handmade quilts over him that the staff reserved for such occasions. The man's face was left visible as per the family's request. The charge nurse arrived to cover for any staff members on the unit who wished to join the family in the procession from the building.

A small chime sounded through the intercom system to alert staff throughout the building that a procession was about to take place. As the staff and funeral home escorts pulled the gurney into the hallway, the chaplain explained that the staff would be honored to join the family in a procession through the hallways and outside to the waiting van, and if they wanted, she would lead them all in a song of their choosing. They welcomed this idea, and in another moment, the whole gathering of twenty-five or thirty people proceeded through the hallways, a hymn echoing behind them, accompanied by sobs of grief and loss.

As the procession reached the sun-soaked pavers just outside the front door of the building, the crowd gathered in a semicircle around the body, and the chaplain expressed how honored the whole staff had been to serve this man and each of his loved ones. The funeral staff wheeled the gurney into the van as the family bid farewell to their loved one. Hugs were plentiful, and it seemed no one was left untouched. "I'm so sorry for your loss," and, "It was such an honor to work with your family," rang out from the many staff members who had gathered at the front walkway. After a few moments of condolences and hugs, the family collected themselves and drove away, while the nurses returned to their patient care.

Point of Intervention: Medical Decision Making

This scene of openly escorting a deceased patient and bereaved family down the hallway and out through the front door of the hospice facility stood in stark contrast to my observations of nurses at the hospital hiding a body as they delivered it to the morgue, or my unexpected discovery of funeral home staff cautiously removing a body through the back door of the hospital. These observations are indicative of the philosophies that guide patient care within each medical model. In conventional cancer care, death is perceived as a failure of treatment, while hospice views death as a natural part of life. Each of these approaches is useful and appropriate at different times along a patient's disease progression.

Early-stage cancers in otherwise healthy individuals respond well to curative treatments, and the conventional philosophy provides the necessary background from which we derive our hopes of remission and survivorship. In advanced illness as disease enters its terminal stages, the hospice philosophy offers patients and their loved ones resources for assessing their goals for how they wish to spend their remaining time, and support for maximizing the quality of that time. When barriers prevent a patient's transition from conventional care into hospice at the time when this shift would provide the most benefit to patients, that is the key point of intervention for maximizing the positive outcomes and minimizing the negative consequences of end-of-life care.

The obstacles to hospice entry span several categories, including health care provider factors, sociodemographics, and financial access.[1] The most apparent barrier that arose in my observations fits within the class of health care provider factors: the lack of open communication about terminal diagnoses, as described in the previous chapter. Other barriers also arose during my data collection, such as the challenge of private insurance not reimbursing hospice care, and cases where terminally ill patients and their families were not ready to forgo the hope of curative treatment. Each of these obstacles deserves consideration if we wish to overcome the challenge of providing appropriate care in line with patient wishes, because the majority of cancer patients report that they wish to die at home. Recent trends suggest that more people are achieving their preferred place of death, but the reality remains that only 42 percent of cancer patients actually die at home, and one quarter continue to die in hospitals.[2] Many of these hospital deaths reportedly occur in the intensive care unit with patients undergoing aggressive, yet futile, medical interventions.

On an individual basis, the misalignment between patient preferences and end-of-life experiences can be distressing to patients, families, and providers. At the population level, this misalignment results in large societal impacts. About one and a half million Americans are newly diagnosed with cancer annually, and more than half a million die of cancer each year. The sheer number of people experiencing an inconsistency between their end-of-life goals and their actual experiences of dying demands that we continue to aspire toward supporting better outcomes for patients and families.

The large number of people who remain within conventional care settings at end of life beyond the time when it is medically appropriate leads to unnecessarily large cumulative environmental and public health impacts. My comparison of the unintended consequences of conventional care, palliative care, and hospice shows that conventional treatments result in far higher ecological impacts than the alternative care models. Conventional care relies on larger volumes of medical supplies and pharmaceuticals and generates larger quantities of waste than palliative care or hospice. As a result, the longer a patient remains within the conventional model beyond the time when it is medically beneficial, the larger the ecological effects of their care. Therefore, my analysis suggests that overcoming the barriers that prevent patients' timely transition into hospice would not only improve patient outcomes, it would also drastically reduce the cumulative ecological impacts of end-of-life care.

By heeding the call for open conversations about death and dying between clinicians and patients, we could realign medical decision making with patient preferences and help patients transition into hospice when medically appropriate, rather than at the last possible moment. Such an achievement has the capacity to benefit patients and families through improved quality of life and bereavement support, while also increasing the sustainability of health care.

Points of Intervention: Clinical Practices

Although my analysis shows that conventional care had the largest ecological impacts of the three medical models, each setting employed some practices that resulted in negative outcomes, and others that helped mitigate these costs. A closer look at the social factors that governed these impacts offers insight for developing the best environmental practices to apply across sites to increase their collective sustainability. These factors tend to center on clinician training (e.g., familiarity with recycling), institutional culture (e.g., donation versus disposal of unused supplies), infrastructure (e.g., space for a medical supply clean area in patient rooms), policy directives (e.g., infection-control standards and drug disposal mandates), and economics (e.g., purchasing of prepackaged kits). In the following sections I present a series of practical implications resulting from my analysis and suggest specific changes that could decrease the ecological impacts of my three research sites. These recommendations could help increase sustainability in other medical settings as well, and may offer general insight to clinicians and hospital administrators seeking to reduce the impacts of the cultural practices employed within their health care institutions.

Medical Waste

- Source reduction: The biggest opportunity for decreasing the impacts of waste is to decrease the overall volume of waste generated. In my observations, a surprisingly large volume of municipal waste comprised unused materials. Having a designated clean area within each patient room was a practice that worked within the hospice facility to allow restocking of unused materials in the supply closet while maintaining compliance with infection control standards.

- Single-stream recycling: To increase the success of recycling within patient areas, staff within purchasing departments could work with staff at recycling facilities to identify and then procure materials that can be recycled via mechanical sorting processes and avoid hand sorting where concerns arise over potential exposure to sharps.
- Staff training: Training protocols that educate all clinical staff about proper disposal practices and potential risks to workers downstream could help protect the health of environmental services staff, waste haulers, and employees at disposal facilities.
- Social connections: Encouraging social connections between clinical staff members, housekeeping, and environmental services staff could further help promote appropriate disposal of waste by promoting a sense of obligation to protect the health of downstream workers.
- Program champions: Identifying respected staff members, particularly nurses, who are champions of recycling and other waste reduction and disposal protocols can increase staff buy-in when new efforts are being implemented.
- Identifying program barriers: Employing a small, targeted trial run of new disposal protocols or programs with a dedicated team can identify and overcome challenges before implementing a large-scale procedural change.
- Technological advancements: Sustainability coordinators could work to identify new technologies that may help reduce the unintended consequences of waste disposal, for example ascertaining options for contracting with waste haulers who use electric vehicles rather than diesel engine trucks to reduce the unintended consequences posed by emissions.
- Infection control: Revisiting infection control policies and practices at the national level could have the largest impact on waste reduction while also maintaining high-quality patient care. I observed several cases where infection control standards led to the

disposal of unused materials still in their packaging because they had entered a patient room. When reconsidering these policies and practices, it is also important to examine the ethics of donating medical supplies that are considered "contaminated" by U.S. standards for redistribution to places around the world experiencing a need for these materials.

Medical Supplies

- Sustainable procurement: As discussed in the introduction to this book, several organizations have helped increase awareness of green purchasing options through group purchasing organizations. Expanding these efforts is necessary to engage each of the 6,000 hospitals in the United States. At the same time, increasing the number of environmentally preferred supplies available and identifying opportunities to incentivize their use will continue to help reduce the unintended consequences of these materials.
- Standardized procedural kits: These kits present a trade-off between the financial cost of supplies and the generation of unnecessary waste. The health care industry may benefit from analyzing cumulative cost savings compared to waste generation, while clinicians could consult with group purchasing organizations to identify shared practices across medical settings that would benefit from having standardized procedural kits. In addition, standardizing practices across institutions could also contribute to minimizing disposal of unused kit materials.
- Reusable versus disposable supplies: Several trade-offs exist between reusable and disposable materials. Short-term economic costs and labor shortages may continue to dictate whether materials are disposed of after a single use, though reassessing the true costs of disposables may tip the balance toward identifying opportunities for reprocessing and reuse.
- Clinical practices: Each health care facility likely has different cultural practices that lead to unnecessary waste and would benefit from identifying points of intervention that will reduce waste while continuing to provide high-quality patient care. The case of medical gloves provides a starting point for examining practices that lead to high volumes of unnecessary waste. In this case, the conventional

practice was to don a new pair of gloves every time a clinician entered a patient room, whereas hospice staff first determined what a patient needed before acquiring a pair of gloves. These two approaches resulted in remarkably different volumes of glove waste.

Pharmaceuticals

- Occupational exposures to chemotherapy: Pre-priming liquid chemotherapy IV bags in the controlled environment of a central pharmacy can help minimize spills, while enforcing requirements for staff to wear appropriate protective equipment when administering chemotherapy can help reduce occupational exposures to these toxic chemicals. The trade-off here is that patients may feel more concerned about their own exposure to chemo when nurses arrive at their bedside in full personal protective equipment, so creating a culture that supports nurses' health is important as well.
- Environmental impacts of hazardous waste incineration: Several liquid medications were available only in certain syringe sizes, which led to a large cumulative volume of waste of both drugs and the plastic syringes. On the conventional and palliative care units, nurses would prepare narcotic medications for a patient by squirting any unneeded amount in a syringe into the hazardous waste container on the unit pharmacy, administer the drug, then dispose of the syringe in the same bin. In comparison, nurses on the hospice inpatient unit maintained a single syringe to administer multiple doses of a medication to a single patient over the course of a single nursing shift. Between doses, the nurse locked the syringe back into the pharmaceutical dispensary. Any unused medication and the syringe were then disposed of at the end of the shift. This prevented the wasting of multiple syringes and unused medication doses for each patient during each shift. On an individual patient level, these concerns may seem trivial, but the reality is that we must account for the cumulative effects of one and a half million patients each year.
- Overprescribing: Many chronically ill patients accumulate large volumes of medications over the course of their care. Policies that restrict overprescribing and overdispensing could help curb the volume of unused drugs, particularly if they allow only a small

volume of medication to be dispensed during a trial period to determine correct dosage and appropriateness for the patient.

- Drug disposal guidelines: Different federal agencies have different goals for pharmaceutical waste disposal. The Drug Enforcement Agency directs health care institutions to dispose of controlled substances by flushing to prevent drug diversion and illegal use, while the Environmental Protection Agency recommends disposal via take-back programs to minimize pharmaceutical pollution in aquatic environments. Meanwhile, nurses providing home health care may want to return unused drugs to take-back programs but do not have the ability to safely transport medications. With the increased availability of drug take-back programs across the country, identifying ways to support returning unused medications to these programs would help decrease the cumulative environmental costs of flushing. Baluster Hospice in particular could develop policies and practices to prevent unused medications from being flushed down the drain, particularly among nurses who work with the in-home hospice program, perhaps by providing small, locked pharmaceutical waste bins that nurses could easily transport between the hospice facility and patients' homes.
- Green pharmacy approaches: Health care institutions could develop policies and position statements in support of green pharmacy approaches that would limit the ecological and public health impacts of drug development, prescription, and disposal. In particular, institutions could release position statements in support of insurance reimbursement practices that would encourage physicians to minimize overprescription of drugs (which may in turn help decrease drug costs).

Barriers to Medically Appropriate Care among Terminally Ill Patients

- Enhancing pathways for patients to transition into hospice when most beneficial: The primary finding of this study was that the barriers that prevent terminally ill cancer patients from entering hospice early enough to reap its benefits also lead to unnecessary ecological impacts. These barriers have been a concern for decades, and as a society we must respond to this challenge to both enhance care for the dying and limit the harms of health care.

The foregoing recommendations provide a starting point for considering the impacts of clinical practices on the broader community. I acknowledge the difficulty in implementing some of these changes, particularly changing nursing practice given the emergent situations and limited time nurses must work within, as well as space limitations for creating clean areas in patient rooms, but such changes would decrease resource waste and potentially lower costs of care. Each of these recommendations deserves further research to identify specific strategies that will be most likely to address environmental impacts while also supporting high-quality patient care.

Conclusion

"Should the environment and public health be given more consideration within end-of-life medicine? Is the medical establishment itself the appropriate place to seek ways for minimizing the environmental impacts of medical practice?"

I asked these questions of several people whom I worked alongside, and all but one responded with a resounding yes, health care should directly consider these concerns. Interestingly, many suggested that both the medical establishment and society in general have a role to play in minimizing the environmental impacts of health care. The one person who disagreed, an administrator at Hopewell's pharmacy department, felt that health care's contribution to our collective environmental challenges was too small to consider in comparison to all the other industries polluting and degrading the environment.

> Obviously we're polluting our environment. Obviously we're doing all kinds of things. And to me to look at medications in the grand scope of things, the impact of that seems so much smaller than everything else. . . . There are so many different chemicals, and they must be having some impact, but what is that impact? How can you measure it? I mean, the quantities of these things are so small you can barely measure any of it, and then when you look at impact, is there really a cause-and-effect relationship between these two things, and if there is, how could you prove it?

This perspective is important because it highlights a general lack of awareness of the known environmental impacts of health care within the industry. As described in the introduction to this book, the health care industry has a significant impact on the environment—particularly regarding climate change and pollution—and several organizations have been working tirelessly to reduce these impacts for decades. My hope is that the analysis this book provides will empower current sustainability efforts to broaden their scope to include the clinical practices and decision making that govern the provision of health care.

Acknowledgments

I have learned so much through this research journey, and I am profoundly grateful to the people who offered me guidance, support, fellowship, and camaraderie along the way. My deepest gratitude goes to the many staff members who allowed me to follow in their footsteps day in and day out, while tirelessly answering my questions and making time for my interviews, even in snowstorms and after working long shifts. Likewise, I am forever grateful to the many patients who allowed me to observe their care. It was a privilege to learn from all these fine people, and I hope that this book will stand as a testament to their experiences and will help enhance the lives of others who follow.

I am grateful for the support offered by many mentors throughout the course of this project. Thanks to Michael Bell, Marty Kanarek, Gregg Mitman, Jonathan Patz, and Laura Senier for their thoughtful guidance that allowed me to venture into this intellectual space with the tools needed to approach such a sensitive topic with humble curiosity. My deepest thanks to Phil Brown for his tremendous capacity as a mentor. In addition, I thank James Cleary, Linda Hogle, Karin Kirchhoff, Jo Scheder, and Frances Westerly for supporting my research and early professional development in countless ways.

Several friends and colleagues have helped me move this project forward, and I especially acknowledge the time and support offered by Aimee Cunningham Walsh, Abby Clark, Peter Boger, Anna Zeide, Trish O'Kane, and Amy Seidl. I also thank the members of my Breadloaf Environmental

Writers Conference workshop, including Stephanie Henck Camillo, William Cocke, Jenna Gersie, Rose Linke, Mila Plavsic, Jess Ryan, Amy Weldon, Isaac Yuen, and our writing mentor, Craig Childs, for their incredible constructive feedback and camaraderie. Gratitude to the anonymous reviewers whose thoughtful suggestions helped guide the completion of this work. My deepest thanks to the editorial staff at Rutgers University Press, and especially Peter Mickulas, for shepherding me through the publication process.

Deepest appreciation for my intellectual community at the Gund Institute for Environment, Larner College of Medicine, and Rubenstein School of Environment and Natural Resources at the University of Vermont, where I have been afforded many opportunities to present various components of this research. I have always walked away stimulated by the interest and curiosity shown by my colleagues.

Gratitude to my sister, Jennifer, and mother, Becky, who were there with me as my father's life ended and this project began, though we did not realize it at the time. And humble thanks to my husband, Brian Sprague, and our son, Ansel, for keeping me grounded in the adventure of living through the many years it has taken to complete this project.

Appendix A

A Note on Methods

My analysis is built on a multi-sited ethnographic approach, which was guided by the work of Marcus (1995) and supported by Savage (2000), who acknowledged that ethnographic methods can provide an in-depth view of beliefs and practices within health settings and help solve problems that other approaches may be unable to address. In addition, I found both the early sociological studies of death and dying completed by Glaser and Strauss (1965) and Sudnow (1967) helpful, both methodologically and analytically.

As described in chapter 1, I determined that a comparative analysis of end-of-life medical care models would offer a unique setting in which to investigate the unintended consequences of medical care and the social factors that govern those outcomes. My research sites included three inpatient units that provided care to terminally ill cancer patients: a conventional cancer inpatient unit, a palliative care unit, and a hospice facility. The conventional and palliative care units were located within a teaching hospital (which I refer to as Hopewell Hospital to protect the identities of my research participants) in a midsized, midwestern U.S. city. The inpatient hospice facility (which I call

Baluster Hospice) had a strong working relationship with the hospital as its primary hospice affiliate and was located a short drive away.

I obtained permission to conduct this research from my university's Institutional Review Board, as well as from administrators at both Hopewell Hospital (director of nursing) and Baluster Hospice (chief operating officer). After obtaining these approvals, I conducted my fieldwork between 2008 and 2010, during which time I spent a total of 255 hours observing the working lives of fifty-four physicians, nurses, housekeepers, and administrators. To better understand my observations, I interviewed seventy-three staff members, including both the people I observed at work and additional staff and administrators. Each person included in this study provided written informed consent to participate. In addition, I obtained institutional records from both facilities to triangulate my ethnographic findings regarding the annual volume of medical supplies used, waste generated, and pharmaceuticals dispensed.

To begin identifying research participants at Hopewell, the nursing administrators invited me to introduce my study at the conventional and palliative care unit monthly nursing staff meetings. In contrast, I reached out directly to physicians and social workers to introduce the study and ask to observe their work. For Environmental Services, the purchasing department, and staff who worked with medical supplies, I contacted departmental supervisors who connected me with staff members to invite into the study. Similarly, administrators at Baluster Hospice preferred that I introduce myself directly to clinicians or contact unit supervisors for permission to invite their staff to participate in the study.

Over the course of time I spent at my field sites, I kept a varied schedule that allowed me to observe a wide range of work shifts and staff members on all days of the week. Although I observed all three work shifts—day, evening, and night—I spent the majority of my time on the day shifts during the regular Monday to Friday workweek. After a short period in the field, I learned that these times corresponded to when the physicians and administrators were on the units making the majority of resource-use decisions.

My observational work consisted of shadowing one staff member at a time, recording the material resources used in that time, along with notes on conversations, interactions, and other pertinent information. In order to shadow staff into patient rooms, I provided each patient with a research information sheet and obtained verbal consent to be present in their room alongside the staff member whom I was observing at the time. At Hopewell

Hospital, I observed the care of fifty-six conventional cancer patients and twenty-one palliative care patients, and at Baluster Hospice I observed the care of fifty-six hospice patients. There were a few rare occasions when patients did not provide consent for me to follow their providers into their room. In these cases, I would record any medical supplies or pharmaceuticals obtained by the nurse I was following, wait in the hall outside the patient's room, and then begin shadowing again once the nurse had completed their task with that patient. There were also a few cases on each unit where staff members, in particular nurses, felt that the patient or family was having a difficult day and was not in a state to have me in the room. In these instances, I erred toward the staff member's caution, and, rather than introducing the study, I simply recorded resources and waited in the hall . For the most part, however, patients were friendly and accommodating to my study and a few were quite interested and struck up conversations about my work—one even handed my information sheet back to me so I could recycle it.

Though I tried to be as unobtrusive as possible and interrupted my participants with questions about what I observed only when we were in times and spaces that seemed appropriate, I was a *participant* observer. This implies that my presence did affect the day-to-day practices of the people I worked alongside, and in turn shaped what I was observing. One clear example of this effect occurred at Hopewell. Once the nursing staff on the conventional unit perceived my work as being "environmental" in nature, they began paying closer attention to waste and decided to start a recycling program within patient rooms as part of their required annual nursing improvement projects. While the implementation of the recycling program can be attributed to my presence and study, it was rather fortuitous since it provided a deeper understanding of the social factors that govern the success or failure of such efforts. In addition, given that both Hopewell Hospital and Baluster Hospice were teaching institutions, the majority of staff members I shadowed were used to having students tag along, which may have helped mitigate the disturbance my presence would have otherwise caused.

Overall, my multi-sited ethnographic approach provided valuable insights related to the question of how medical decision making drives resource use within clinical settings, especially since my three sites shared several staff members. The conventional cancer unit and palliative care unit were located on a shared hallway where they also shared a supply closet, unit pharmacy, nursing staff, and administrators. I often shadowed nurses and administrators between the two units on any given day, and they offered invaluable thoughts on the similarities and differences in care between these two sites. Likewise, several pharmacists and physicians shared their time between Hopewell Hospital and Baluster Hospice, and they provided instrumental insights by comparing their experiences with patient care and clinical decision making within these facilities.

Appendix B

A Note on Theory

Two primary questions guided this research: (1) What are the unintended environmental and public health consequences that arise from clinical practices and medical decision making? and (2) What are the social factors that govern these outcomes? To answer these questions I employed the sociological *theory of environmental flows* to help illuminate the many factors inherent within the sustainability of a system.[1] In a way, the theory reflects life cycle analysis, but rather than quantifying the resources used in the cradle to grave production of a material, environmental flows theory suggests that to identify points of intervention for overcoming sustainability challenges, we must look beyond the material to the social structures that govern, or have power over, the movement of materials.

Environmental flows theory builds on the more general *sociology of flows* described by Manuel Castells and John Urry, who argued that modern borderless societies require a new kind of "sociology beyond societies." They suggested that modern life is constituted through movement across social boundaries and is governed by multinational corporations and other entities beyond the control of the traditional nation-state. The environmental sociologists Arthur Mol and Gert Spaargaren found this approach to be quite applicable to modern environmental issues, and they offered environmental flows theory as a way to examine both the movement of natural resources and the social factors governing that movement, including political economy, institutional forces, and cultural factors.

Typically, sociologists use environmental flows theory to examine the governance of materials from cradle to grave. Since the practice of medicine relies heavily on material inputs to provide patient care—utilizing everything from diagnostic equipment to pharmaceuticals to hospital beds—it makes sense that examining the material flows of medical supply chains from their sources to their eventual sinks in the natural world would provide useful information about the sustainability of medical care. Through the following pages, I offer an examination of material flows through my research sites, along with four theoretical contributions to environmental flows theory. I argue that these additions to the theory help reveal points of intervention that can lead to more sustainable outcomes in systems, including global supply chains of medical supplies and pharmaceuticals, and the movement of patients through medical care settings.

First, while the movement of materials may be central to analyzing environmental flows, it is also necessary to acknowledge the importance of the movement of people and information relative to material flows. This realization led me to uncover three dimensions of environmental flows: the *material dimension* (the physical object), the *information dimension* (knowledge about the processes and content that make up environmental flows), and the *identity dimension* (the relational aspect of environmental flows to people and places).

Second, while observing the handling of medical waste streams I became aware of factors that prevent staff members from being able to see occupational health and safety hazards. From this insight, I offer the idea that *visibility* and *invisibility* of the three dimensions of environmental flows help explain certain aspects of their movement.

Third, although environmental flows theory originally focused on studying movement, sometimes things do not move. Here, I build on the term *immobilities* to reflect how motionlessness relates to sustainability.[2]

Finally, I suggest that when examining points of intervention within a system, we must consider the *flexibility* or *rigidity* (i.e., the likelihood of changing) of the factors that govern flows in order to lead to better outcomes. Together, examining the governance of all three dimensions of environmental flows (material, information, identity), their visibilities/invisibilities, their mobilities/immobilities, and the flexibility/rigidity of the factors that govern their movement, provides us with a more robust opportunity for identifying opportunities that can help increase the sustainability of a system.

Dimensions of Environmental Flows

Environmental flows theory to date has focused on the movement and governance of materials. In my three research sites—conventional, palliative care, and hospice inpatient units—many materials are constantly moving to provide patient care. However, from my collective observations across these three research sites, I began to see the connections between the movement of materials and the movement of people and information. Twenty-four hours a day, seven days a week, people—nurses, doctors, students, administrators, support staff, visitors, and patients—are moving through these places. They carry with them information of various types—diagnoses, prognoses, procedures, options, goals—and interact with information in various formats—electronic, interpersonal, educational. People, information, and materials are all important parts of the story.

Consider for a moment an IV bag. The bag has, most obviously, a *material dimension*—it is a physical object; you can touch it, it is made of plastic. Less obvious is the bag's *information dimension*—what it contains, its therapeutic properties, the functions it serves, how to properly dispose of it after use. Less obvious still is the bag's *identity dimension*—the location where it originated, the hands it passed through before arriving at the IV pole where it hangs, the specific nurse who administered it to a specific patient. The IV bag has a relationship to the people and places from which it was created and to which it will travel when discarded.

Each of these dimensions are themselves governed by a number of factors that together determine the sustainability of the "system" of IV bag production, use, and disposal. The availability of petroleum that the plastic originated from is dictated by global production and market demand as well as multinational corporations, not to mention U.S. military policies. Knowledge of its contents—its information dimension as represented by the label detailing what it contains—is governed by national and institutional safety standards (e.g., an IV bag holding a hazardous chemotherapy drug must display an orange warning sticker to alert staff of potential exposures). The identity dimension of the bag (and what it contains) is determined by international trade agreements, the availability of workers in the plastics manufacturing industry at home and abroad, and union contracts of the truckers who transport it from the warehouse to the hospital's loading dock, to name a few.

Together, these three dimensions and the factors that govern them add complexity to our understanding of the sustainability of environmental flows, but in ways that allow us to better examine points of intervention that may lead to decreased ecological and public health impacts of supply chains.

Visibility of the Dimensions of Environmental Flows

While standing on the loading dock, peering into municipal garbage bags with Alice, a member of the Environmental Services department at the hospital, I began to see the importance of visibility of the different dimensions of environmental flows. For example, the black garbage bag makes it impossible for the materials inside to be visible, thereby potentially putting workers at risk if they are unknowingly handling infectious waste. Similarly, the separation between job duties among clinical workers and Environmental Services staff make these people invisible to one another and increases the possibility that infectious waste will enter the municipal waste stream. Identifying strategic ways to decrease problems of visibility may help decrease the negative impacts of material resource use.

Some invisibility is necessary given the global scale of these supply chains—not everyone throughout the entire chain can know or be known by everyone else. However, employing strategic approaches to provide knowledge about the people and places most likely to be affected by the practices used in one location could help minimize the cumulative effects of medical supply chains. Research is needed to identify specific approaches that might work, but one possible solution may be to provide information about the people, places, and processes immediately upstream and downstream from each point in the system to build a sense of responsibility and connection between workers. Such strategies could be tested to determine whether increasing the visibility of relationships within the system could increase behaviors in support of occupational health across the supply chain.

We must also consider that there are times when information and identities need to remain invisible in order for the system to function properly. The Health Insurance Portability and Accountability Act offers a good example. If personal medical records became visible, medical institutions would be legally responsible for the breach in confidentiality, and a number of negative outcomes would result (such as paying for damages or a possible loss of accreditation), so organizations put a number of protections in place to maintain the invisibility of patient records. This example suggests that we must carefully consider the visibility of dimensions throughout the system under scrutiny to ensure appropriate outcomes.

Mobility of the Dimensions of Environmental Flows

Likewise, in order to support the sustainability of a given system we must carefully determine the impact of mobility within each dimension of flows. In the case of pathogens, employing infection control standards that prevent the mobility of microbes leads to desired outcomes. In contrast, when social factors prevent the movement of information and identity dimensions of medical supply chains, clinical staff may be more likely to waste these materials. Here, I build on the concept of *immobility*, which takes into account the fact that sometimes flows stop moving (Adey, 2006). In the case of medical supplies, though, I suggest that only certain dimensions—identities and information—of the environmental flows of supplies are immobile, rather than the entire flow.

As I stood in the supply closets at Hopewell Hospital and Baluster Hospice, the majority of medical supplies that surrounded me provided extremely limited visibility of their identity and information dimensions. There were no pictures on the boxes identifying their landscapes of origin or the workers who had manufactured the products. There were no labels telling me the distance the items had traveled, their ecological footprints, or the total cost of their cradle-to-grave process. The identity and information dimensions of these materials were invisible, not simply because they were inside a black bag, they were invisible because they never even made it to the loading dock. Even the manufacturing personnel I contacted to inquire about some of these aspects of products did not know where the products had originated. Although the material dimensions of supply flows are mobile, many aspects of their identity and information dimensions do not move—they are immobile.

The proper mobilization/immobilization of the dimensions of environmental flows decreases downstream impacts—as in the case of infection control limiting the spread of pathogens. In contrast, when the identities of manufacturing workers and information about poor working conditions in global supply chains are immobilized, institutions continue to demand and use products in unsustainable ways. Therefore, there are times when mobility leads to poor outcomes, and other times when immobility leads to negative impacts—but note that in both examples provided, immobility typically leads to higher material use. Careful examination of various environmental flows is necessary to ensure that policy and practice recommendations consider the ecological and public health outcomes of a specific resource.

Flexibility of Factors Governing Environmental Flows

The governance of each of the three dimensions of environmental flows shapes the system and helps define the environmental and public health outcomes of their combined movement. Vast corporate policies and international trade agreements currently separate the people and places where products originate from those where supplies are used (and where waste is disposed of). To increase sustainability across supply chains, we need to uncover places within the governance of these systems that are flexible enough to change in a way that would enhance the visibility and mobility of identities. Here, I offer the opposing terms of *flexibility* and *rigidity* to provide some context for how we might strategically select points of intervention within the governance of environmental flows that determine the sustainability of systems.

A flexible governance structure permits change; a rigid structure does not. Minimizing the impacts of medical care on the broader community requires governance factors to balance strategically between flexibility and rigidity. If governance factors are too flexible, individual actors may be left to devise their own practices, resulting in high environmental costs. We see this clearly in the case of pharmaceutical waste disposal practices at Baluster Hospice. Since drug disposal guidelines from federal agencies do not align with institutional accrediting agencies, medical facilities must determine their own methods for drug disposal. As a result, Baluster trained nursing staff to dispose of all unused medications down the drain, thereby posing risks to the aquatic environment.

In contrast, there are times when governance can be too rigid, which again leads to negative outcomes. An example of such rigidity is the institutional policy at Hopewell Hospital that created the palliative care service. By declaring palliative care a consult service, rather than a service offered concurrently alongside conventional cancer care, the program was stymied in its ability to provide counseling to patients throughout their course of treatment that would help people with life-limiting illness to prioritize their end-of-life goals. The rigidity of this system meant that the palliative care team was often called in for consultation very late in patient care, thereby leading many patients to remain in material-based conventional settings longer than medically beneficial. This unnecessary length of stay within the conventional model added to the cumulative impacts of care on sustainability outcomes.

Baluster Hospice's approach to minimizing medical supply waste offers one example of flexible governance that was able to enhance sustainability. In this setting, nurses received training to limit the number of supplies they brought into patient rooms. As the program unfolded, follow-up materials included information about the volume of materials diverted from the trash, along with the cost savings created by the program. This information helped increase the local visibility of information that in turn helped decrease the environmental impact of waste. The ability for such changes to occur suggests that governance factors at Baluster Hospice were flexible enough to allow for change in response to emerging needs. For example, the institutional factors that governed the facility's infection control practices responded to the need for creating "clean" areas in each patient room, as well as to the need for staff training to reduce supply waste.

Interactions between Visibility, Mobility, and Governance

Finally, we must recognize interactions between mobility, visibility, and governance factors. Otherwise, unintended consequences of changing one of the three aspects without considering the others may actually be worse than the original environmental and public health impacts. The clearest example of the importance of considering all three factors arose through my observations of the failed recycling program at Hopewell's conventional and palliative care units. When members of the nursing staff implemented the new recycling program on the units, they focused the campaign solely on the mobility of information—specifically, they developed a list of materials that could be recycled, and they shared that information with other nurses through staff meetings and through a few signs posted around the unit. After a few months, the unacceptable amount of nonrecyclable materials entering the recycling stream forced Hopewell to discontinue all recycling on inpatient units and nearly caused the hospital to lose its contract with their recycling hauler—certainly a situation that could have been considered worse than the original problem.

The discontinuation of the inpatient unit recycling program resulted from overlooked governance and visibility factors. In terms of governance, a number of people beyond the nursing staff had access to the waste bin near the doors of patient rooms, so a universal education program was needed to ensure that others did not toss trash into the recycling bin. The lack of a shared avenue for information exchange across all staff, patients, and visitors proved to be a major challenge to the program. In terms of visibility, downstream workers at the recycling processing plant had concerns about their own safety when they began seeing medical tubing showing up on the sorting conveyor belts. By overlooking the importance of recycling facility workers to the success of the program, the nurses who implemented recycling on the units compromised their own goal of waste reduction. As this evidence suggests, all three aspects of flows—mobility, visibility, and governance—interact in important ways that determine the sustainability of systems. By broadening our understanding of environmental flows and their governance beyond the purely material, we advance new opportunities to enhance sustainability.

Appendix C

Institutional Data on Materials Used at Hopewell Hospital and Baluster Hospice

Table C.1

Comparison of the most commonly used medical supplies at Hopewell Hospital's conventional cancer and palliative care units (combined) and Baluster Hospice's acute care inpatient unit

	Total	Total per patient day
Hopewell Hospital Conventional Cancer and Palliative Care Units		
Medical gloves	663,596	67 (33 pairs)
Wound dressings (4×4)	47,850	4.80
Plastic one-ounce medicine cups	21,850	2.2
Intravenous tubing and supplies	24,434	2.46
Intravenous solution bags	18,201	2.00
Baluster Hospice Facility		
Medical gloves	450,000	2.04 (1 pair)
Oral swabs	180,000	0.82
Incontinence briefs	90,176	0.41
Wound dressings (3×3 and 4×4)	80,352	0.36
Protective skin cream	10,512	0.05

NOTE: Since the conventional cancer and palliative care units at Hopewell function as one cost center institutional data on medical supply usage were only available for their combined totals. Note that a "patient day" represents the total number of days that the total number of patients received care from the organization (e.g., 100 patients in a hospital for 1 day equals 100 patient days), and provides an easier way to compare data.

Table C.2
The ten most commonly dispensed pharmaceuticals on the conventional care, palliative care, and hospice inpatient units (as per institutional data)

Drug (purpose)	Total number of doses dispensed	Doses per patient day
Hopewell Hospital Conventional Cancer Unit		
Acyclovir (antiviral)	9,608	1.23
Pantoprazole (stomach acid reduction)	6,050	0.78
Oxycodone (pain—Schedule II controlled substance)	5,251	0.67
Acetaminophen (pain/fever reduction)	5,004	0.64
Ondansetron (antinausea)	5,003	0.64
Morphine (pain—Schedule II controlled substance)	4,601	0.59
Lorazepam (sedative/muscle relaxant)	4,362	0.56
Cefepime (antibiotic)	3,736	0.48
Dexamethasone (steroid/immunosuppressant)	3,471	0.45
Docusate-senna (laxative)	3,118	0.40
Hopewell Hospital Palliative Care Unit		
Morphine (pain—Schedule II controlled substance)	1,609	1.61
Lorazepam (sedative/muscle relaxant)	655	0.66
Acetaminophen (pain/fever reduction)	495	0.50
Docusate-senna (laxative)	458	0.46
Oxycodone (pain—Schedule II controlled substance)	413	0.41
Hydromorphone (pain—Schedule II controlled substance)	397	0.40
Dexamethasone (steroid/immunosuppressant)	367	0.37
Pantoprazole (stomach acid reduction)	317	0.32
Ondansetron (antinausea)	259	0.26
Gabapentin (pain/neuropathic pain)	194	0.19
Baluster Hospice Acute Care Inpatient Unit		
Morphine (pain—Schedule II controlled substance)	15,231	2.33
Lorazepam (sedative/muscle relaxant)	6,930	1.06
Docusate-senna (laxative)	4,864	0.74
Albuterol (chronic obstructive pulmonary disease)	4,095	0.63
Acetaminophen (pain)	3,827	0.58
Ibuprofen (pain)	3,079	0.47
Hydromorphone (pain—Schedule II controlled substance)	2,914	0.45
Gabapentin (pain/neuropathic pain)	2,891	0.44
Carbidopa-levodopa (Parkinson's disease)	2,543	0.39
Haloperidol (antipsychotic/delirium)	2,510	0.38

NOTE: The data for Hopewell's conventional cancer unit include pharmaceuticals dispensed to patients on the hematology, bone marrow transplant, and oncology services, combined.

Notes

Introduction

1 Andrew Jameton and Jessica Pierce, bioethicists at the University of Nebraska, were among the first to identify the cumulative impacts of health care on the environment (Jameton, 1999; Jameton & Pierce, 2001).
2 Despite the ongoing Covid-19 pandemic, climate change is broadly considered the greatest threat to human health (Wise, 2021). The health care sector accounts for up to 5 percent of global and 10 percent of U.S. greenhouse gas emissions (Eckelman et al., 2020; Eckelman & Sherman, 2016; Tennison et al., 2021). These statistics have prompted a call to action within the health care industry to curb these emissions in order to better support health outcomes (Dzau et al., 2021).
3 U.S. Global Change Research Program (2016).
4 The Commonwealth Fund provides analysis of several health-related outcomes compared to health care spending among OECD countries (Tikkanen & Abrams, 2020). The thirty-six member countries of the OECD include the following high-income countries: Australia, Canada, France, Germany, the Netherlands, New Zealand, Norway, Sweden, Switzerland, the United Kingdom, and the United States. Estimates show that health care spending in the United States outranks all peer OECD nations, although Americans have the shortest life expectancy (Roser, 2017).
5 Health Care Without Harm (https://noharm.org/) reports a number of successes from their work to reduce the environmental footprint of health care (Health Care Without Harm, 2022).
6 Practice Greenhealth was originally named Hospitals for a Healthy Environment (H2E) but was renamed in 2008 (https://practicegreenhealth.org/our-history).
7 Frumkin and Coussens (2007).
8 Complete agendas of Yale's workshops on health care sustainability are available at https://ysph.yale.edu/yale-center-on-climate-change-and-health/healthcare-sustainability-and-public-health/sustainability-health-care-symposium/.

9 Cost savings are a key driver in health care greening, with estimates that implementing sustainable practices, including energy use reduction, waste reduction, and more efficient supply purchasing in hospitals nationwide, could save USD 5.4 billion over a five-year period, and up to USD 15 billion over ten years (Kaplan et al., 2012). Meanwhile, patient satisfaction has been described as one of the key social factors in the greening of health care (Faezipour & Ferreira, 2013).

10 The ethical implications of incorporating environmental concerns into medical decision making must be considered in any actions taken to improve the sustainability of the health care system. Here, I point to the works of Jameton, Pierce, and Carrick to ensure the consideration of patient autonomy within efforts to support planetary health (Carrick, 1999, 2005, 2010; Jameton, 1999; Jameton & Pierce, 2001; Pierce & Jameton, 2005).

Chapter 1 Focal Point

1 The names of all people and places described in this book have been changed to protect the confidentiality of my research subjects.

2 Some of the most pivotal books on the sociology of death and dying include *On Death and Dying* by Elisabeth Kubler-Ross, and *How We Die* by Sherwin Nuland. Others have remarkable titles, including *If You Want to Know If You're Dying, Ask the Cleaning Lady, and Other Thoughts on Life and Happiness* by Sally Peterson Karioth.

3 Thirty percent of Medicare dollars are spent on the 5 percent of recipients who die each year (Davis et al., 2016). Furthermore, in the United States, the average annual health care expense is USD 10,739 per person, which accounts for 18 percent of the gross domestic product (Centers for Medicare & Medicaid Services, 2018) As chronic illness sets in (here, estimated as an average cost over the final three years of life), the cost rises to USD 51,799 per year. The cost of the final year of life jumps drastically to USD 80,094 ((French et al., 2017).

4 Cancer is second only to heart disease in the leading causes of death in the United States, with 609,640 people dying of cancer in 2017, or about 1,670 deaths each day (American Cancer Society, 2018). Current estimates suggest that 38.4 percent of Americans will be diagnosed with some form of cancer in their lifetime.

5 The differing philosophies and goals of care for conventional curative care (Gawande, 2014), palliative care (Elsayem et al., 2004), and hospice (Saunders et al., 1981) help explain many of my observations regarding the way death is treated within these three settings.

6 Throughout Western history, hospitals and hospices—from the Latin *hospes* (host)—were places of last resort where travelers, refugees, and the poor could find food and shelter—rather than medical care—usually provided by religious orders (Risse, 1999). The secularization of English hospitals occurred during Henry VIII's English Reformation in the sixteenth century, and in the seventeenth century hospitals became a place for military veterans and the working poor to receive medical care. Some hospitals that had previously been affiliated with religious institutions now reopened as secular facilities, and new voluntary hospitals began opening as well. The first American hospitals included Pennsylvania Hospital (1751), New York Hospital (1791), and Massachusetts General Hospital (1821), which followed the model of English voluntary hospitals by

providing care to the indigent (Vogel, 1980). For further reading on the history of medicine and the dying, I recommend the work of Ariès (2013) and Starr (1982).

7 A review of patient preferences for cancer care found that the ability to receive treatment at home was one of the top reasons why patients choose oral over IV cancer drugs (Eek et al., 2016).

8 The National Cancer Institute provides five-year survival rates for all common cancer sites at https://seer.cancer.gov/statfacts/ (National Cancer Institute, 2019).

9 Aggressive cancer treatments can cause harm to patients, particularly as they near the end of life (Chen et al., 2016; Pope, 2018). Aggressive, curative cancer treatments that have been documented within the last year of life include radiation treatment (Tseng et al., 2018) and pharmaceuticals (Davis, 2015), particularly chemotherapy (Harrington & Smith, 2008). Furthermore, a recent study of 681 cancer patients showed that within the last year of life, 30 percent of patients experienced two emergency department visits, 17 percent spent more than thirty days in the hospital, and 38 percent died in the hospital (Henson et al., 2016). The term "futile" in the context of end-of-life care is very loaded within the health care community. However, interviews with clinicians have helped clarify and define the term as being, "failure to achieve goals of care that offer a benefit to the patient's quality of life and are proportionate to the risks, harms and costs" (Jox et al., 2012).

10 The Institute of Medicine created expert panels in 1997 and again in 2014 to explore the treatment of terminally ill patients in America (Field & Cassel, 1997; Institute of Medicine, 2014). Their reports called for the creation and support of palliative care options in health care systems (Tulsky, 2015).

11 Several studies have indicated the lower cost of palliative care as compared to conventional health care; Smith et al. (2014) offer a broad review of this literature.

12 Centers for Medicare and Medicaid Services, 2019.

13 The number of people who prefer to die at home versus the number who actually do has been studied extensively. A systematic review of the literature suggests that the majority of people would prefer to die at home supported by palliative care and hospice models, and that this preference tends to remain constant throughout disease progression (Gomes et al., 2013). Among cancer patients, the number of people who would reportedly prefer to die at home ranges from 49 to 90 percent (Gomes & Higginson, 2006; Higginson & Sen-Gupta, 2000). In reality, 42 percent of cancer patients achieve a home death, and this number varies across demographics, with clear differences by race and gender (Chino et al., 2018). A systematic review of twenty-three peer-reviewed academic articles that provide a clear overview of several research methodologies was used to elicit these data (Nilsson et al., 2017).

14 According to the most recent data available from National Hospice and Palliative Care Organization, the average length of stay for patients in hospice in 2016 was seventy-one days. Just under one-third of hospice patients died within seven days of enrolling in the program, and another 25 percent died within a month (National Hospice and Palliative Care Organization, 2018).

Chapter 2 Medical Waste

1 A review of life-cycle analysis studies for how best to manage paper waste determined that recycling is the best choice as compared to incineration or

disposal in a landfill (Villanueva & Wenzel, 2007). In addition, the EPA archives describe the benefits of recycling paper as compared to landfilling or incineration (Environmental Protection Agency, 2016c). More recently, the EPA offers evidence of the economic benefits of recycling, in general, stating that the recycling industry provides more than 750,000 jobs, nearly $37 billion in wages, and $7 billion in tax revenue annually (Environmental Protection Agency, 2018a, 2018b).

2 *Pulp and Paper Industry: Chemicals* provides an overview of the chemicals used in the papermaking and recycling processes, which includes both caustic agents and heavy metals, such as mercury (Bajpai, 2015). Occupational exposures to these chemicals can cause ill health to workers, though evidence of such exposures is not readily available.

3 While 4.5 pounds is the average amount of municipal solid waste produced per person in the United States each day, only about 53 percent of this waste is landfilled, whereas the remainder is recycled (26 percent), composted (9 percent), or used in waste-to-energy incinerators (13 percent) (Environmental Protection Agency, 2018a, 2018b).

4 From my observations on the three inpatient units, the majority of municipal trash produced on the conventional cancer unit consisted of medical supplies and their packaging from each patient room—polyvinyl chloride bags from saline and nonhazardous drugs, IV tubing, IV dressing kits, needleless syringes, gloves, plastic pill cups and packaging, emesis basins, Foley catheters, hot packs, drink pitchers and their Styrofoam liners. The municipal trash on the palliative care unit contained materials similar to those on the conventional cancer unit—supplies and packaging—although I observed less IV use on the palliative care unit (0.7 IV bags used per palliative care patient compared to 2.4 bags per patient in the conventional services), which corresponded to less IV bag waste. In addition, since the target length of stay for patients on the palliative care unit was three days (compared to three days for oncology, eleven days for hematology, and twenty days for bone marrow transplant) these patients typically did not require changes in IV tubing and IV dressings (infection control standards require these to be changed every three days), and therefore had less waste from these supplies. At the hospice inpatient unit, gloves, pill cups and packaging, paper towels (staff members at Baluster washed their hands with soap and water more frequently than at Hopewell, where staff members preferred to use hand sanitizer), and disposable "chucks" underpads were the majority of municipal trash items that I observed.

5 The average fuel economy of diesel-powered garbage trucks is estimated at between 2 and 3 miles per gallon (Sandhu et al., 2016). Diesel emissions are of major concern to public health and have been associated with a number of respiratory ailments, including airway inflammation, decreased lung function, oxidative stress, and asthma (Brunekreef et al., 1997; Ghio et al., 2012; Ristovski et al., 2012; Sydbom et al., 2001). Children living near roads with heavy truck traffic are particularly vulnerable to the effects of diesel air pollutants since their respiratory systems are still developing (Liu & Grigg, 2018; Van Vliet et al., 1997). Furthermore, greenhouse gas emissions from long-distance transport threaten public health via a number of climate change–related outcomes, including increased risk of injury from flooding and extreme weather events, increased risk of malnutrition caused by drought and crop failure, and increased incidence of emerging

infectious and vectorborne diseases (Haines et al., 2006; Levy & Patz, 2015; Patz et al., 2005).

6 Landfill emissions are approximately 50 percent methane and 50 percent carbon dioxide. Both of these compounds are potent greenhouse gases, but evidence shows that methane is roughly thirty times more potent than carbon dioxide at trapping heat in earth's atmosphere (Environmental Protection Agency, 2019a).

7 The future consequences of failed landfill facilities raise intergenerational concerns (El-Fadel et al., 1997; Naveen et al., 2017).

8 Several studies are useful when considering the public health impacts of landfills (Banzhaf et al., 2019; Bullard, 2000; Dolk et al., 1998; Kim & Williams, 2016; Vrijheid, 2000, 2009).

9 There is a long history of racism in the context of who hauls the trash (Pellow, 2002; Zimring, 2017). Current national statistics show that waste management and remediation services are performed by a diversity of people, including 77 percent Caucasian, 18 percent Black or African American, and 2 percent Asian, with 23 percent identifying as Hispanic or Latino (U.S. Department of Labor, Bureau of Labor Statistics, 2019). In comparison, the national demographics of all hospital staff combined include 73 percent Caucasian, 15 percent Black or African American, 9 percent Asian, and 11 percent Hispanic or Latino.

10 The most recent national mandate, adopted in January 2009, sets the minimum wage at $7.25 per hour (U.S. Department of Labor, 2019).

11 Environmental Protection Agency (2019c).

12 Francis et al. (1997).

13 Esposito (1997).

14 Environmental Protection Agency (2019c).

15 Hospital-acquired infections are among the top ten leading causes of death in the United States (Haque et al., 2018). The cost of treating hospital-acquired infections can range up to USD 45 billion per year (Scott, 2009).

16 Estimated costs of hospital-acquired infections depend on the type of infection and range from $25,000 to $45,000 per patient (Arefian et al., 2016; Pittet et al., 1994). The cost savings of implementing infection control practices average over $13,000 per month.

17 Limiting the amount of materials being brought into patient rooms is an approach that fits in well with source reduction programs (Tchobanoglous & Kreith, 2002).

18 Recent literature has begun to offer some insight into the challenges of donating medical supplies and equipment to resource-poor areas of the world. For example, in The Gambia it became apparent that donated oxygen concentrators—equipment that helps provide a steady flow of oxygen to a patient through an oxygen mask—were breaking down because the hospital had neither staff with appropriate technical expertise to maintain them nor the proper electrical voltage to use the concentrators (Howie et al., 2008). Similarly, ethical concerns are implicated in the case of donated pacemakers, which, once removed from a patient in the developed world, are sterilized and then surgically implanted into patients in the developing world (Kirkpatrick et al., 2010).

19 Several concerns arise from the donation of medical materials to places that need these goods but may not have the ability to maintain them (Bhadelia, 2016; World Health Organization, 2000).

Chapter 3 Medical Supplies

1 The question of whether clinicians should consider the environmental impacts of medical care was prompted by Andrew Jameton, a bioethicist (Jameton, 1999). The qualitative life-cycle analysis of latex gloves was completed by Pierce and Kerby (1999).

2 The world's militaries are considered to be the largest polluting force on Earth (Clark & Jorgenson, 2012). Military installations and weapons testing destroy native flora and fauna, and leave toxic substances—lead, mercury, arsenic, solvents, pesticides, and radioactive waste—in their wake. Land mines left behind after armed conflict has ended prove deadly to wildlife as well as farmers and their livestock. Above all, the U.S. military is the "single largest producer of greenhouse gases in the world" (Crawford, 2019). These emissions lead to climate change, the most pressing public health challenge of our time, even in the face of the COVID-19 pandemic (Fearnow, 2020; Gates, 2020).

3 The annual economic value of these ecosystem services within Malaysia's six marine parks alone is estimated at 8.7 billion Malaysian Ringgit (USD 2 billion) (Barbier, 2017; Malaysia, 2020). In contrast, offshore oil and gas drilling accounts for USD 71 billion, or 20 percent of Malaysia's gross domestic product (Mustafa, 2011).

4 Malaysian oil spill incidents are governed by a variety of complex socio-political factors (Ishak et al., 2021). The International Tanker Owners Pollution Federation insurance company provides limited details about the 1997 major oil spill off the coast of Malaysia on their website at https://www.itopf.org/knowledge-resources/countries-territories-regions/countries/malaysia/. Research on the impact of oil spills on the marine ecosystem in the region included the *Evoikos* spill (Jaswar & Maimun, 2014), though limited information on the environmental impacts of this spill are possible because of the lack of baseline data on marine biodiversity in the area (Tan et al., 1999).

5 Several occupational health risks occur within the nitrile industry (Chaiear, 2001; Dotson et al., 2016; Santos-Burgoa et al., 1992).

6 Skokan (2021).

7 Ayliffe (2000).

8 Two historians, Gerald Markowitz and David Rosner, provide a revelatory overview of the public health risks of the plastics industry in their book, *Deceit and Denial: The Deadly Politics of Industrial Pollution* (Markowitz & Rosner, 2002). An excellent overview of mercury bioaccumulation is available at https://www.canada.ca/en/environment-climate-change/services/pollutants/mercury-environment/health-concerns/food-chain.html. For information about mercury advisories in fish, see https://www.epa.gov/mercury/guidelines-eating-fish-contain-mercury. The World Health Organization provides an overview of the health implications of dioxins at https://www.who.int/news-room/fact-sheets/detail/dioxins-and-their-effects-on-human-health. Details of the Seveso disaster are available through the *New York Times* archive at https://www.nytimes.com/1976/08/19/archives/seveso-disaster.html and in academic articles (Homberger, Reggiani, Sambeth, & Wipf, 1979). Several follow-up studies have chronicled the long-term cancer risks and other health effects of the Seveso disaster (Bertazzi et al., 2001; Pesatori et al., 2003).

9 Research on the health risks of vinyl chloride have been ongoing since the 1970s (Kielhorn et al., 2000; Makk et al., 1974; Mundt et al., 2017). The fascinating histories of towns along Cancer Alley that have been moved to reduce environmental exposure are chronicled in *Deceit and Denial* (Markowitz & Rosner, 2002).
10 Resources for understanding the effects of DEHP include Health Care Without Harm (https://noharm-uscanada.org/issues/us-canada/phthalates-and-dehp), the U.S. Food and Drug Administration (https://www.fda.gov/media/114001/download), and several academic articles, including Mallow & Fox (2014).
11 For alternatives to PVC and DEHP, see https://noharm-uscanada.org/content/europe/pvc-free-alternatives.
12 The Lowell Center for Sustainable Production maintains a searchable database (http://www.chemicalspolicy.org/chemicalspolicy.us.state.database.php) that provides a list of all state and regional chemical policies related to sustainable practices in health care, such as purchasing and disposal. In the region where Hopewell and Baluster are located, bans and restrictions on the purchase and disposal of mercury-containing thermometers and a general law promoting pollution prevention through the use of nonhazardous materials could be applicable, though not enforceable, at my research sites.
13 See table C.1 in Appendix C for data on per patient usage of medical gloves.

Chapter 4 Pharmaceuticals

1 Data on trends in global consumption of pharmaceuticals are available through the IMS Institute for Healthcare Informatics (Informatics, 2015). In the United States, the most recent data available from the Centers for Disease Control and Prevention show that 46 percent of Americans have used at least one prescription drug in the past thirty days (Martin, 2019).
2 Bioprospecting has helped to identify numerous natural compounds that offer useful therapies for many human ailments, including cancer, inflammation, pain, and infection, but the practices used to harvest wild products can be devastating to their native ecosystems (Sen & Samanta, 2014). Likewise, when indigenous knowledge of the medicinal properties of natural products is taken from communities and exploited for drug development without fair, equitable, and mutually agreed upon compensation, the issue of biopiracy can threaten the livelihood of indigenous people (Efferth et al., 2016).
3 There are several excellent resources that describe the story of Taxol (Joyce, 1993; Kaiser, 2001; Walsh & Goodman, 1999; Weaver, 2014).
4 In 2015, the global greenhouse gas emissions (measured as metric tons of carbon dioxide equivalent per million USD revenue) were 48.55 for the pharmaceutical industry, compared with 31.5 for the automotive industry (Belkhir & Elmeligi, 2019). The energy needed to produce pharmaceuticals is the major driver of greenhouse gas emissions from the industry (Wernet et al., 2010). The EPA Energy Star program partnered with researchers at the University of California at Berkeley to develop a guide for energy efficiency within the pharmaceutical industry (Galitsky, 2008). Since its beginning, twenty-one drug companies have joined the program to reduce energy use and associated costs (Environmental Protection Agency, 2020).

5 Limited data is available on the total proportion of drugs consumed by Americans that originate outside the United States; the estimate of 80 percent of drugs originating in China and India comes from a statement by U.S. senator Chuck Grassley (R-Iowa) regarding quality control for pharmaceuticals and active ingredients entering the United States from abroad (U.S. Senate Committee on Finance, 2019). In India, interest in adopting sustainable practices has been expressed but will take time to adopt (Chaturvedi et al., 2017). In China, total greenhouse gas emissions from the pharmaceutical industry rose from 23.03 metric tons in 2000 to 55.34 metric tons in 2016 (Gao et al., 2019). One comprehensive study of pollution emitted from major Chinese industries, including pharmaceuticals, shows how the burning of coal contributes to several ill-health effects (Hu et al., 2019).

6 Each of these pollutants has been researched extensively, including particulate matter (World Health Organization, 2018), nitrous oxides (Environmental Protection Agency, 2016a), sulfur dioxide (Environmental Protection Agency, 2019b), and volatile organic compounds. One recent study specifically investigated the release of these chemicals from drug manufacturing facilities in China, where an estimated 80 percent of medications used in the United States are produced (Hu et al., 2019).

7 A description of the common side effects of chemotherapy are available through the American Cancer Society (American Cancer Society, 2020).

8 The National Institute for Occupational Safety and Health (NIOSH) maintains a list of antineoplastic drugs that are categorized as "hazardous" (National Institute for Occupational Safety and Health, 2016).

9 Limited research has examined the potential for exposure to oral chemotherapy drugs. One study that looked at exposure among at-home caregivers who dispensed antineoplastic pills to pediatric cancer patients found that 90 percent of respondents did not use recommended precautions to limit their exposure to the medication (Held et al., 2013).

10 The health risks posed by occupational exposure to antineoplastic chemicals are well-documented (National Institute for Occupational Safety and Health, 2017), with robust reviews available on the effects of chemotherapy exposure on workers' reproductive health (Connor et al., 2014).

11 The 1,814 nurses who participated in this survey were predominantly female (96 percent), white (92 percent), aged 41 years or older (83 percent), and had more than ten years of experience administering antineoplastic drugs (60 percent) (DeJoy et al., 2017). The survey asked health care workers about their behaviors in accordance with the National Institute for Occupational Safety and Health guidelines for safe handling of antineoplastic chemicals (National Institute for Occupational Safety and Health, 2004, 2008).

12 Baluster Hospice and Hopewell Hospital contracted with the same waste hauler to transport hazardous waste for incineration, and so both facilities contributed to the environmental and public health outcomes of incineration (Environmental Protection Agency, 2016b). However, they produced drastically different volumes of hazardous waste: Hopewell created 74 tons of hazardous waste for incineration annually, whereas Baluster had accumulated such a small amount over a two-year period that they had not yet contacted the waste hauler to come and pick it up.

13 Domingo et al. provide a robust review of current research on the health effects of hazardous waste incinerators (Domingo et al., 2020). Tait et al. offer a review of health effects related to waste incinerators in general, and provide additional evidence of reproductive effects, including low sperm counts, miscarriage, preterm birth, and congenital birth defects, along with increased risk of developmental delay in children. In addition, these authors stated that "there is insufficient evidence to conclude that any incinerator is safe" (Tait et al., 2020).

14 There is a broad history of environmental injustice in relation to the siting of hazardous facilities, beginning with the classic works of Robert Bullard and the United Church of Christ (Bullard, 2000a; United Church of Christ Commission for Racial Justice, 1987).

15 According to the 2010 U.S. Census, the population of Port Arthur, Texas, included a total of 53,818 people, 40 percent of whom were Black or African American, 29.6 percent were Hispanic or Latinx, and 23.4 percent lived below the poverty line (Prochaska et al., 2014). In addition, the rate of unemployment in Port Arthur is historically almost double that of the rest of Texas.

16 Robert Bullard is again credited with identifying the case of environmental justice in Emelle, Alabama (Bullard, 2000b). Despite several attempts by the local community and national groups to close the landfill, especially since it was the primary site for disposal of materials from Superfund cleanups across the country, it remains open as of 2022 (Hess & Satcher, 2019).

17 According to the CDC, 69,029 people died of drug overdose in the one-year period ending in February 2019 (Centers for Disease Control and Prevention, 2019).

18 In several recent studies, flushing has been reported as a commonly used disposal method among the general population, including 28 percent of survey respondents in California (Kotchen et al., 2009), 26 percent in Southern California (Law et al., 2015), and 12 percent in Vermont (Vatovec, 2021). Drugs can accumulate and require disposal for many reasons—patients stop using them or doctors direct them to discontinue use, a greater volume is dispensed than required, the drugs expire—all of which point to concerns of overprescribing, overpurchasing, and overdispensing (Ruhoy & Daughton, 2008; Vatovec et al., 2017, 2021).

19 Pharmaceuticals have been detected globally in surface waters, such as lakes and streams, beginning with the U.S. Geological Survey national reconnaissance (Kolpin et al., 2002) and more recently in studies conducted around the world (Deo, 2014; Fekadu et al., 2019). These contaminants have also been found in groundwater (Fram & Belitz, 2011; Zainab et al., 2020), estuaries (Letsinger et al., 2019), and marine environments (Ojemaye & Petrik, 2019). Since many cities draw their drinking water from these sources, it is not surprising that these pollutants can also be detected in drinking water, but what might be surprising is that these chemicals are not filtered out through municipal drinking water filtration and so can be found in treated drinking water (Benotti et al., 2009; Furlong et al., 2017; Khan & Nicell, 2015).

20 Many drugs are considered to have "pseudo-persistence" in the environment since they are continually introduced through wastewater, and as such they are cause for concern regarding the health of aquatic species and ecosystems (Tijani et al., 2016). A number of studies provide evidence of the effects of pharmaceutical

pollutants on aquatic species (Arnold et al., 2014; Bringolf et al., 2010; Brodin et al., 2014; Brooks, 2014; Fursdon et al., 2019; Klimaszyk & Rzymski, 2017; Sarma et al., 2017; Sathishkumar et al., 2020; Valenti, Jr. et al., 2012). Endocrine-disrupting chemicals, including birth control pills, can reduce the reproductive success of wild fish by up to 76 percent (Harris et al., 2011). Anticancer drugs, including tamoxifen, are of particular concern because they target DNA and other cellular processes that promote growth, and as a result can have genetic effects in nontarget species (Fonseca et al., 2019; Heath et al., 2016).

21 A study comparing the use of antianxiety and antidepressant medications among more than 3,000 cancer survivors and more than 44,000 adults with no cancer history found that those with a history of cancer were nearly twice as likely to take these medications: 16.8 percent of cancer survivors took antianxiety drugs compared to 8.6 percent of the general public; 14.1 percent versus 7.8 percent for antidepressants; and 19.1 percent versus 10.4 percent for both (Hawkins et al., 2017). Several studies have examined the effects of fluoxetine (Prozac) on wildlife, with studies showing several toxic effects, including the inhibition of normal growth (Brooks et al., 2003; Duarte et al., 2020) and changes in aggressive behaviors among exposed fish (Perreault et al., 2003).

22 The federal agencies that offer guidance about drug disposal provide conflicting recommendations. The FDA highlights the "flush list" (U.S. Food and Drug Administration, 2020), whereas the EPA recommends not flushing drugs unless specifically directed to do so by the medication label or patient information (Environmental Protection Agency, 2011). The Drug Enforcement Administration website provides links that direct consumers to both of these other agencies' directives (U.S. Department of Justice Drug Enforcement Administration, 2020).

Chapter 5 Patients

1 Beginning palliative care earlier has several benefits for patients, including increased length of survival and higher quality of life (Temel et al., 2010), and higher likelihood of having their end-of-life care preferences met (Mack et al., 2010).

2 Hospice reportedly provides better outcomes for families of the deceased, who report higher quality of care provided to their loved one as compared to conventional care (Kumar et al., 2017; Teno et al., 2004; Wright et al., 2016), and longer length of life among bereaved spouses (Christakis & Iwashyna, 2003).

3 Several benefits of hospice extend to hospitals that help increase their profitability (Harrison et al., 2005).

4 The National Hospice and Palliative Care Organization provides utilization statistics in their annual reports (National Hospice and Palliative Care Organization, 2020), and recent research provides further insight into hospice underutilization (Cagle et al., 2020). I calculated these numbers based on the following information from the National Hospice and Palliative Care Organization: 600,000 people die of cancer each year in the United States, 30 percent of whom never enroll in hospice.

5 A study of over 2,100 patients with end-stage cancer found that 24 percent of patients received chemotherapy within one month of death (Pacetti et al., 2015). It is important to note that a portion of these cases may be attributed to hospice

itself as some hospice organizations allow chemotherapy if it is necessary to reduce symptoms and improve quality of life for a patient.

6 Prognosis presents a number of challenges to physicians (Fox et al., 1999; Knaus et al., 1995).

7 Aksoy et al. (2016).

Chapter 6 Conclusions and Practical Implications

1 Barriers to hospice care vary between urban and rural areas and across demographics and include a range of socioeconomic and cultural factors (Parajuli, Tark, Jao, & Hupcey, 2020).

2 See note 13 in chapter 1 for literature detailing patient preferences versus location of death. Approximately 12.5 percent of cancer deaths occur in the intensive care unit (Aksoy et al., 2016).

Appendix B A Note on Theory

1 The sociology of flows is the work of Manuel Castells and John Urry (Castells, 1996, 2004; Urry, 2000, 2002, 2003) who inspired the environmental sociologists Arthur Mol and Gert Spaargaren to develop the theory of environmental flows (Mol & Spaargaren, 2003). The earlier work on natural resource "additions and withdrawals" is attributed to Schnaiberg (Schnaiberg, 1980). In addition, the work of Foucault is useful here when applying the concept of the power of social structures in governance (Foucault, 1982).

2 Here I build on the term "immobility" (Adey, 2006). I suggest that the immobility of flows is an important consideration when identifying points of intervention in a system.

References

Adey, P. (2006). If mobility is everything then it is nothing: Towards a relational politics of (im)mobilities. *Mobilities, 1*(1), 75–94. https://doi.org/10.1080/17450100500489080.

Aksoy, Y., Kaydu, A., Sahin, O. F., & Kacar, C. K. (2016). Analysis of cancer patients admitted to intensive care unit. *Northern Clinics of Istanbul, 3*(3), 217.

American Cancer Society. (2018). Cancer statistics. https://www.cancer.gov/about-cancer/understanding/statistics.

American Cancer Society. (2020). Chemotherapy side effects. https://www.cancer.org/treatment/treatments-and-side-effects/treatment-types/chemotherapy/chemotherapy-side-effects.html.

Arefian, H., Vogel, M., Kwetkat, A., & Hartmann, M. (2016). Economic evaluation of interventions for prevention of hospital acquired infections: A systematic review. *PLOS ONE, 11*(1), e0146381.

Ariès, P. (2013). *The hour of our death.* Vintage.

Arnold, K. E., Brown, A. R., Ankley, G. T., & Sumpter, J. P. (2014). Medicating the environment: assessing risks of pharmaceuticals to wildlife and ecosystems. *Philosophical Transactions of the Royal Society B: Biological Sciences, 369*(1656), 20130569.

Ayliffe, G. A. J., Fraise, A. P., Geddes, A. M., & Mitchell, K. (2000). *Control of hospital infection: A practical handbook* (4th ed.). Arnold.

Bajpai, P. (2015). *Pulp and paper industry: Chemicals.* Elsevier.

Banzhaf, S., Ma, L., & Timmins, C. (2019). Environmental justice: The economics of race, place, and pollution. *Journal of Economic Perspectives, 33*(1), 185–208.

Barbier, E. B. (2017). Marine ecosystem services. *Current Biology, 27*(11), R507–R510.

Belkhir, L., & Elmeligi, A. (2019). Carbon footprint of the global pharmaceutical industry and relative impact of its major players. *Journal of Cleaner Production, 214*, 185–194.

Benotti, M. J., Trenholm, R. A., Vanderford, B. J., Holady, J. C., Stanford, B. D., & Snyder, S. A. (2009). Pharmaceuticals and endocrine disrupting compounds in US drinking water. *Environmental Science & Technology, 43*(3), 597–603.

Bertazzi, P. A., Consonni, D., Bachetti, S., Rubagotti, M., Baccarelli, A., Zocchetti, C., & Pesatori, A. C. (2001). Health effects of dioxin exposure: A 20-year mortality study. *American Journal of Epidemiology, 153*(11), 1031–1044.

Bhadelia, N. (2016). Rage against the busted medical machines. National Public Radio, *Goats and Soda*. https://www.npr.org/sections/goatsandsoda/2016/09/08/492842274/rage-against-the-busted-medical-machines.

Bringolf, R. B., Heltsley, R. M., Newton, T. J., Eads, C. B., Fraley, S. J., Shea, D., & Cope, W. G. (2010). Environmental occurrence and reproductive effects of the pharmaceutical fluoxetine in native freshwater mussels. *Environmental Toxicology and Chemistry, 29*(6), 1311–1318.

Brodin, T., Piovano, S., Fick, J., Klaminder, J., Heynen, M., & Jonsson, M. (2014). Ecological effects of pharmaceuticals in aquatic systems—impacts through behavioural alterations. *Philosophical Transactions of the Royal Society B: Biological Sciences, 369*(1656), 20130580.

Brooks, B. W. (2014). Fish on Prozac (and Zoloft): Ten years later. *Aquatic Toxicology, 151*, 61–67. https://doi.org/10.1016/j.aquatox.2014.01.007.

Brooks, B. W., Foran, C. M., Richards, S. M., Weston, J., Turner, P. K., Stanley, J. K., Solomon, K. R., Slattery, M., & La Point, T. W. (2003). Aquatic ecotoxicology of fluoxetine. *Toxicology Letters, 142*(3), 169–183.

Brunekreef, B., Janssen, N. A., de Hartog, J., Harssema, H., Knape, M., & van Vliet, P. (1997). Air pollution from truck traffic and lung function in children living near motorways. *Epidemiology, 8*(3), 298–303.

Bullard, R. D. (Ed.). (2000a). *Dumping in Dixie: Race, class, and environmental quality* (3rd ed.). Westview.

Bullard, R. D. (2000b). Environmental racism in the Alabama Blackbelt. http://www.ejrc.cau.ehttp://www.ejrc.cau.edu/envracismalablackbelt.htm du/envracismalablackbelt.htm.

Cagle, J. G., Lee, J., Ornstein, K. A., & Guralnik, J. M. (2020). Hospice utilization in the United States: A prospective cohort study comparing cancer and noncancer deaths. *Journal of the American Geriatrics Society, 68*(4), 783–793.

Carrick, P. (1999). Environmental ethics and medical ethics: Some implications for end-of-life care, Part II. *Cambridge Quarterly of Healthcare Ethics, 8*(2), 250–256.

Carrick, P. (2005). The hidden costs of environmentally responsible health care. *Perspectives in Biology and Medicine, 48*(3), 453–458.

Carrick, P. (2010). Deep ecology and end-of-life care. In J. Pierce & G. Randels (Eds.), *Contemporary bioethics: A reader with cases* (pp. 704–713). Oxford University Press.

Castells, M. (1996). *The rise of the network society*. Blackwell.

Castells, M. (2004). *The network society: A cross-cultural perspective*. Edward Elgar.

Centers for Disease Control and Prevention. (2019). NCHS releases new monthly provisional estimates on drug overdose deaths. https://www.cdc.gov/nchs/pressroom/podcasts/20190911/20190911.htm.

Centers for Medicare and Medicaid Services. (2018). National health expenditure data: Historical. https://www.cms.gov/research-statistics-data-and-systems/statistics-trends-and-reports/nationalhealthexpenddata/nationalhealthaccounts historical.html.

Centers for Medicare and Medicaid Services. (2019). Hospice. https://www.cms.gov/Medicare/Medicare-Fee-for-Service-Payment/Hospice/index.html.

Chaiear, N. (2001). *Health and safety in the rubber industry* (Vol. 138). Smithers Rapra Publishing.

Chaturvedi, U., Sharma, M., Dangayach, G., & Sarkar, P. (2017). Evolution and adoption of sustainable practices in the pharmaceutical industry: An overview with an Indian perspective. *Journal of Cleaner Production, 168*, 1358–1369.

Chen, R. C., Falchook, A. D., Tian, F., Basak, R., Hanson, L., Selvam, N., & Dusetzina, S. (2016). Aggressive care at the end-of-life for younger patients with cancer: Impact of ASCO's Choosing Wisely campaign. *Journal of Clinical Oncology 34*(18, suppl). https://doi.org/10.1200/JCO.2016.34.18_suppl.LBA10033.

Chino, F., Kamal, A. H., Leblanc, T. W., Zafar, S. Y., Suneja, G., & Chino, J. P. (2018). Place of death for patients with cancer in the United States, 1999 through 2015: Racial, age, and geographic disparities. *Cancer, 124*(22), 4408–4419.

Christakis, N. A., & Iwashyna, T. J. (2003). The health impact of health care on families: A matched cohort study of hospice use by decedents and mortality outcomes in surviving, widowed spouses. *Social Science & Medicine, 57*(3), 465–475.

Clark, B., & Jorgenson, A. K. (2012). The treadmill of destruction and the environmental impacts of militaries. *Sociology Compass, 6*(7), 557–569.

Connor, T. H., Lawson, C. C., Polovich, M., & McDiarmid, M. A. (2014). Reproductive health risks associated with occupational exposures to antineoplastic drugs in health care settings: A review of the evidence. *Journal of Occupational and Environmental Medicine/American College of Occupational and Environmental Medicine, 56*(9), 901.

Crawford, N. C. (2019). *Pentagon fuel use, climate change, and the costs of war.* Watson Institute, Brown University.

Davis, C. (2015). Drugs, cancer and end-of-life care: A case study of pharmaceuticalization? *Social Science & Medicine, 131*, 207–214.

Davis, M. A., Nallamothu, B. K., Banerjee, M., & Bynum, J. P. (2016). Patterns of healthcare spending in the last year of life. *Health Affairs (Project Hope), 35*(7), 1316.

DeJoy, D. M., Smith, T. D., Woldu, H., Dyal, M.-A., Steege, A. L., & Boiano, J. M. (2017). Effects of organizational safety practices and perceived safety climate on PPE usage, engineering controls, and adverse events involving liquid antineoplastic drugs among nurses. *Journal of Occupational and Environmental Hygiene, 14*(7), 485–493.

Deo, R. P. (2014). Pharmaceuticals in the surface water of the USA: A review. *Current Environmental Health Reports, 1*(2), 113–122.

Dolk, H., Vrijheid, M., Armstrong, B., Abramsky, L., Bianchi, F., Garne, E., Nelen, V., Robert, E., Scott, J. E., Stone, D., & Tenconi, R. (1998). Risk of congenital anomalies near hazardous-waste landfill sites in Europe: The EUROHAZCON study. *The Lancet, 352*(9126), 423–427. https://doi.org/10.1016/s0140-6736(98)01352-x.

Domingo, J. L., Marquès, M., Mari, M., & Schuhmacher, M. (2020). Adverse health effects for populations living near waste incinerators with special attention to hazardous waste incinerators: A review of the scientific literature. *Environmental Research, 187*, 109631.

Dotson, G. S., Maier, A., Parker, A., & Haber, L. T. (2016). Immediately dangerous to life or health (IDLH) value profile: Acrylonitrile (CAS no. 107-13-1). National

Institute for Occupational Safety and Health. Education and Information Division. https://stacks.cdc.gov/view/cdc/41593.
Duarte, I. A., Reis-Santos, P., Novais, S. C., Rato, L. D., Lemos, M. F., Freitas, A., Pouca, A. S. V., Barbosa, J., Cabral, H. N., & Fonseca, V. F. (2020). Depressed, hypertense and sore: Long-term effects of fluoxetine, propranolol and diclofenac exposure in a top predator fish. *Science of the Total Environment, 712*, 136564.
Dzau, V. J., Levine, R., Barrett, G., & Witty, A. (2021). Decarbonizing the US health sector—a call to action. *New England Journal of Medicine, 385*(23), 2117–2119.
Eckelman, M. J., Huang, K., Lagasse, R., Senay, E., Dubrow, R., & Sherman, J. D. (2020). Health care pollution and public health damage in the United States: An update: Study examines health care pollution and public health damage in the United States. *Health Affairs (Millwood), 39*(12), 2071–2079.
Eckelman, M. J., & Sherman, J. (2016). Environmental impacts of the US health care system and effects on public health. *PLOS ONE, 11*(6), e0157014.
Eek, D., Krohe, M., Mazar, I., Horsfield, A., Pompilus, F., Friebe, R., & Shields, A. L. (2016). Patient-reported preferences for oral versus intravenous administration for the treatment of cancer: A review of the literature. *Patient Preference and Adherence, 10*, 1609. https://doi.org/ 10.2147/PPA.S106629.
Efferth, T., Banerjee, M., Paul, N. W., Abdelfatah, S., Arend, J., Elhassan, G., Hamdoun, S., Hamm, R., Hong, C., Kadioglu, O., Naß, J., Ochwangi, D., Ooko, E., Ozenver, N., Saeed, M. E. M., Schneider, M., Seo, E. J., Wu, C. F., Yan, G. . . . Kadioglu, O. (2016). Biopiracy of natural products and good bioprospecting practice. *Phytomedicine, 23*(2), 166–173.
El-Fadel, M., Findikakis, A. N., & Leckie, J. O. (1997). Environmental Impacts of Solid Waste Landfilling. *Journal of Environmental Management, 50*(1), 1–25. https://doi.org/10.1006/jema.1995.0131.
Elsayem, A., Swint, K., Fisch, M. J., Palmer, J. L., Reddy, S., Walker, P., Zhukovsky, D., Knight, P., & Bruera, E. (2004). Palliative care inpatient service in a comprehensive cancer center: Clinical and financial outcomes. *Journal of Clinical Oncology, 22*(10), 2008–2014. https://doi.org/10.1200/JCO.2004.11.003.
Environmental Protection Agency. (2011). How to dispose of medicines properly. https://archive.epa.gov/region02/capp/web/pdf/ppcpflyer.pdf.
Environmental Protection Agency. (2016a, September 8, 2016). Basic Information about NO2. https://www.epa.gov/no2-pollution/basic-information-about-no2.
Environmental Protection Agency. (2016b). Medical waste frequent questions. https://archive.epa.gov/epawaste/nonhaz/industrial/medical/web/html/mwfaqs.html.
Environmental Protection Agency. (2016c). Wastes—resource conservation—common wastes & materials—paper recycling. https://archive.epa.gov/wastes/conserve/materials/paper/web/html/index-2.html#benefits.
Environmental Protection Agency. (2018a). National overview: Facts and figures on materials, wastes and recycling. https://www.epa.gov/facts-and-figures-about-materials-waste-and-recycling/national-overview-facts-and-figures-materials.
Environmental Protection Agency. (2018b). Recycling basics. https://www.epa.gov/recycle/recycling-basics.
Environmental Protection Agency. (2019a). Basic information about landfill gas. https://www.epa.gov/lmop/basic-information-about-landfill-gas.

Environmental Protection Agency. (2019b). Sulfur dioxide basics. https://www.epa.gov/so2-pollution/sulfur-dioxide-basics.

Environmental Protection Agency. (2019c). Medical waste. https://www.epa.gov/rcra/medical-waste.

Environmental Protection Agency. (2020). Energy Star focus on energy efficiency in pharmaceutical manufacturing. https://www.energystar.gov/industrial_plants/measure-track-and-benchmark/energy-star-energy-11.

Esposito, K. (1997). Operating with a new attitude—*Wisconsin Natural Resources* magazine—October 1997. http://dnr.wi.gov/wnrmag/html/stories/1997/oct97/medwaste.htm.

Faezipour, M., & Ferreira, S. (2013). A system dynamics perspective of patient satisfaction in healthcare. *Procedia Computer Science, 16*, 148–156. https://doi.org/10.1016/j.procs.2013.01.016.

Fearnow, B. (2020). U.N. secretary-general says climate change devastation will be "many times greater" than coronavirus pandemic. https://www.newsweek.com/un-secretary-general-says-climate-change-devastation-will-many-times-greater-coronavirus-1499304.

Fekadu, S., Alemayehu, E., Dewil, R., & Van der Bruggen, B. (2019). Pharmaceuticals in freshwater aquatic environments: A comparison of the African and European challenge. *Science of the Total Environment, 654*, 324–337.

Field, M. J., & Cassel, C. K. (Eds.). (1997). *Approaching death: Improving care at the end of life/Committee on Care at the End of Life, Division of Health Care Services, Institute of Medicine*. National Academies Press.

Fonseca, T., Carriço, T., Fernandes, E., Abessa, D., Tavares, A., & Bebianno, M. (2019). Impacts of in vivo and in vitro exposures to tamoxifen: Comparative effects on human cells and marine organisms. *Environment International, 129*, 256–272.

Foucault, M. (1982). The subject and power. *Critical Inquiry, 8*(4), 777–795.

Fox, E., Landrum-McNiff, K., Zhong, Z., Dawson, N. V., Wu, A. W., & Lynn, J. (1999). Evaluation of prognostic criteria for determining hospice eligibility in patients with advanced lung, heart, or liver disease. *JAMA, Journal of the American Medical Association, 282*(17), 1638.

Fram, M. S., & Belitz, K. (2011). Occurrence and concentrations of pharmaceutical compounds in groundwater used for public drinking-water supply in California. *Science of the Total Environment, 409*(18), 3409–3417.

Francis, M. C., Metoyer, L. A., & Kaye, A. D. (1997). Exclusion of noninfectious medical waste from the contaminated waste stream. *Infection Control and Hospital Epidemiology, 18*(9), 656.

French, E. B., McCauley, J., Aragon, M., Bakx, P., Chalkley, M., Chen, S. H., Christensen, B. J., Chuang, H., Côté-Sergent, A., De Nardi, M., & De Nardi, M. (2017). End-of-life medical spending in last twelve months of life is lower than previously reported. *Health Affairs (Millwood), 36*(7), 1211–1217.

Frumkin, H., & Coussens, C. (2007). *Green healthcare institutions: Health, environment, and economics*. National Academies Press.

Furlong, E. T., Batt, A. L., Glassmeyer, S. T., Noriega, M. C., Kolpin, D. W., Mash, H., & Schenck, K. M. (2017). Nationwide reconnaissance of contaminants of emerging concern in source and treated drinking waters of the United States: Pharmaceuticals. *Science of the Total Environment, 579*, 1629–1642.

Fursdon, J. B., Martin, J. M., Bertram, M. G., Lehtonen, T. K., & Wong, B. B. (2019). The pharmaceutical pollutant fluoxetine alters reproductive behaviour in a fish independent of predation risk. *Science of the Total Environment, 650*, 642–652.

Galitsky, C., Chang, S., Worrell, E., & Masanet, E. (2008). Energy efficiency improvement and cost saving opportunities for the pharmaceutical industry: An Energy Star guide for energy and plant managers. Ernest Orlando Lawrence Berkeley National Laboratory, University of California. https://escholarship.org/uc/item/9zw158vm.

Gao, Z., Geng, Y., Wu, R., Chen, W., Wu, F., & Tian, X. (2019). Analysis of energy-related CO2 emissions in China's pharmaceutical industry and its driving forces. *Journal of Cleaner Production, 223*, 94–108.

Gates, B. (2020). COVID-19 is awful: Climate change could be worse. https://www.gatesnotes.com/Energy/Climate-and-COVID-19.

Gawande, A. (2014). *Being mortal: Medicine and what matters in the end*. Metropolitan Books.

Ghio, A. J., Smith, C. B., & Madden, M. C. (2012). Diesel exhaust particles and airway inflammation. *Current Opinion in Pulmonary Medicine, 18*(2), 144–150.

Glaser, B. G., & Strauss, A. L. (1965). *Awareness of dying*. Aldine.

Gomes, B., Calanzani, N., Gysels, M., Hall, S., & Higginson, I. J. (2013). Heterogeneity and changes in preferences for dying at home: A systematic review. *BMC Palliative Care, 12*(1), 1–13.

Gomes, B., & Higginson, I. J. (2006). Factors influencing death at home in terminally ill patients with cancer: Systematic review. *BMJ, 332*(7540), 515–521. https://doi.org/10.1136/bmj.38740.614954.55.

Haines, A., Kovats, R. S., Campbell-Lendrum, D., & Corvalán, C. (2006). Climate change and human health: Impacts, vulnerability and public health. *Public Health, 120*(7), 585–596.

Haque, M., Sartelli, M., McKimm, J., & Bakar, M. A. (2018). Health care–associated infections—an overview. *Infection and Drug Resistance, 11*, 2321. https://doi.org/10.2147/IDR.S177247.

Harrington, S. E., & Smith, T. J. (2008). The role of chemotherapy at the end of life: "When is enough, enough?" *Journal of the American Medical Association, 299*(22), 2667–2678.

Harris, C. A., Hamilton, P. B., Runnalls, T. J., Vinciotti, V., Henshaw, A., Hodgson, D., Coe, T. S., Jobling, S., Tyler, C. R., & Sumpter, J. P. (2011). The consequences of feminization in breeding groups of wild fish. *Environmental Health Perspectives, 119*(3), 306–311.

Harrison, J. P., Ford, D., & Wilson, K. (2005). The impact of hospice programs on U.S. hospitals. *Nursing Economics, 23*(2), 78.

Hawkins, N. A., Soman, A., Lunsford, N. B., Leadbetter, S., & Rodriguez, J. L. (2017). Use of medications for treating anxiety and depression in cancer survivors in the United States. *Journal of Clinical Oncology, 35*(1), 78.

Health Care Without Harm. (2022). History and victories. https://noharm-uscanada.org/content/us-canada/history-and-victories.

Heath, E., Filipič, M., Kosjek, T., & Isidori, M. (2016). *Fate and effects of the residues of anticancer drugs in the environment*. Springer.

Held, K., Ryan, R., Champion, J. M., August, K., & Radhi, M. A. (2013). Caregiver survey results related to handling of oral chemotherapy for pediatric patients with

acute lymphoblastic leukemia. *Journal of Pediatric Hematology/Oncology, 35*(6), e249–e253.
Henson, L. A., Gomes, B., Koffman, J., Daveson, B. A., Higginson, I. J., & Gao, W. (2016). Factors associated with aggressive end of life cancer care. *Supportive Care in Cancer, 24*(3), 1079–1089.
Hess, D. J., & Satcher, L. A. (2019). Conditions for successful environmental justice mobilizations: An analysis of 50 cases. *Environmental Politics, 28*(4), 663–684.
Higginson, I. J., & Sen-Gupta, G. J. (2000). Place of care in advanced cancer: A qualitative systematic literature review of patient preferences. *Journal of Palliative Medicine, 3*(3), 287.
Homberger, E., Reggiani, G., Sambeth, J., & Wipf, H. (1979). The Seveso accident: Its nature, extent and consequences. *Annals of Occupational Hygiene, 22*(4), 327–370.
Howie, S. R., Hill, S. E., Peel, D., Sanneh, M., Njie, M., Hill, P. C., Mulholland, K., & Adegbola, R. A. (2008). Beyond good intentions: Lessons on equipment donation from an African hospital. *Bulletin of the World Health Organization, 86*, 52–56.
Hu, Y., Li, Z., Wang, L., Zhu, H., Chen, L., Guo, X., An, C., Jiang, Y., & Liu, A. (2019). Emission factors of NOx, SO2, PM and VOCs in pharmaceuticals, brick and food industries in Shanxi, China. *Aerosol and Air Quality Research, 19*(8), 1784–1797.
IMS Institute for Healthcare Informatics. (2015). Global medicines use in 2020: Outlook and implications. https://www.iqvia.com/-/media/iqvia/pdfs/institute-reports/global-medicines-use-in-2020.
Institute of Medicine. 2014. *Dying in America: Improving quality and honoring individual preferences near the end of life*. National Academies Press.
Ishak, I. C., Arof, A. M., Zoolfakar, M. R., Nizam, A. S., & Jainal, N. (2021). The challenges of the oil spill preparedness and responses. *Advanced Engineering for Processes and Technologies II, 147*, 59–65.
Jameton, A. (1999). Conflicts between individual health and nature preservation. *Cambridge Quarterly of Healthcare Ethics, 8*(1), 97–98.
Jameton, A., & Pierce, J. (2001). Environment and health: 8: Sustainable health care and emerging ethical responsibilities. *Canadian Medical Association Journal, 164*(3), 365–369.
Jaswar, M. R., & Maimun, A. (2014). Effect of oil spill pollution in Malacca Strait to marine ecosystem. *Latest Trends in Renewable Energy and Environmental Informatics*, 373–377. ISBN: 978-1-61804-175-3.
Jox, R. J., Schaider, A., Marckmann, G., & Borasio, G. D. (2012). Medical futility at the end of life: The perspectives of intensive care and palliative care clinicians. *Journal of Medical Ethics, 38*(9), 540–545.
Joyce, C. (1993). Taxol: Search for a cancer drug. *Bioscience, 43*(3), 133.
Kaiser, C. (2001). Book review [Review of the book *The story of Taxol: Nature and politics in the pursuit of an anti-cancer drug*, by Jordan Goodman and Vivien Walsh]. *Journal of Medicinal Chemistry, 44*(20), 3335–3336. https://doi.org/10.1021/jm010248e.
Kaplan, S., Sadler, B., Little, K., Franz, C., & Orris, P. (2012). *Can sustainable hospitals help bend the health care cost curve?* Commonwealth Fund.
Khan, U., & Nicell, J. (2015). Human health relevance of pharmaceutically active compounds in drinking water. *AAPS Journal, 17*(3), 558–585.

Kielhorn, J., Melber, C., Wahnschaffe, U., Aitio, A., & Mangelsdorf, I. (2000). Vinyl chloride: Still a cause for concern. *Environmental Health Perspectives, 108*(7), 579–588.

Kim, M. A., & Williams, K. A. (2016). Lead levels in landfill areas and childhood exposure: An integrative review. *Public Health Nursing, 34*(1), 87–97.

Kirkpatrick, J. N., Papini, C., Baman, T. S., Khota, K., Eagle, K. A., Verdino, R. J., & Caplan, A. L. (2010). Reuse of pacemakers and defibrillators in developing countries: Logistical, legal, and ethical barriers and solutions. *Heart Rhythm, 7*(11), 1623–1627.

Klimaszyk, P., & Rzymski, P. (2018). Water and aquatic fauna on drugs: What are the impacts of pharmaceutical pollution? In M. Zelenakova (Ed.), *Water management and the environment: Case studies* (pp. 255–278). Springer.

Knaus, W. A., Harrell, F. E., Jr., Lynn, J., Goldman, L., Phillips, R. S., Connors, A. F., Dawson, N. V., Fulkerson, W. J., Califf, R.M., Desbiens, N., Layde, P., & Wagner, D. P. (1995). The SUPPORT prognostic model: Objective estimates of survival for seriously ill hospitalized adults. *Annals of Internal Medicine, 122*(3), 191.

Kolpin, D. W., Furlong, E. T., Meyer, M. T., Thurman, E. M., Zaugg, S. D., Barber, L. B., & Buxton, H. T. (2002). Pharmaceuticals, hormones, and other organic wastewater contaminants in US streams, 1999–2000: A national reconnaissance. *Environmental Science & Technology, 36*(6), 1202–1211. https://doi.org/10.1021/Es011055j.

Kotchen, M., Kallaos, J., Wheeler, K., Wong, C., & Zahller, M. (2009). Pharmaceuticals in wastewater: Behavior, preferences, and willingness to pay for a disposal program. *Journal of Environmental Management, 90*(3), 1476–1482. https://doi.org/10.1016/j.jenvman.2008.10.002.

Kumar, P., Wright, A. A., Hatfield, L. A., Temel, J. S., & Keating, N. L. (2017). Family perspectives on hospice care experiences of patients with cancer. *Journal of Clinical Oncology, 35*(4), 432.

Law, A. V., Sakharkar, P., Zargarzadeh, A., Tai, B. W. B., Hess, K., Hata, M., Mireles, R., Ha, C., & Park, T. J. (2015). Taking stock of medication wastage: Unused medications in US households. *Research in Social and Administrative Pharmacy, 11*(4), 571–578. http://ac.els-cdn.com/S1551741114003337/1-s2.0-S1551741114003337-main.pdf?_tid=50c28464-f850-11e6-946b-00000aacb361&acdnat=1487693631_342d85b1a7bcc034f8ca15c0d2aa4a19.

Letsinger, S., Kay, P., Rodríguez-Mozaz, S., Villagrassa, M., Barceló, D., & Rotchell, J. M. (2019). Spatial and temporal occurrence of pharmaceuticals in UK estuaries. *Science of the Total Environment, 678*, 74–84.

Levy, B., & Patz, J. (2015). *Climate change and public health*. Oxford University Press.

Liu, N. M., & Grigg, J. (2018). Diesel, children and respiratory disease. *BMJ Paediatrics Open, 2*(1), 1–8. https://doi.org/10.1136/bmjpo-2017-000210.

Mack, J. W., Weeks, J. C., Wright, A. A., Block, S. D., & Prigerson, H. G. (2010). End-of-life discussions, goal attainment, and distress at the end of life: Predictors and outcomes of receipt of care consistent with preferences. *Journal of Clinical Oncology, 28*(7), 1203.

Makk, L., Creech, J. L., Whelan, J. G., & Johnson, M. N. (1974). Liver damage and angiosarcoma in vinyl chloride workers: A systematic detection program. *JAMA, 230*(1), 64–68.

Malaysia, J. T. L. (2020). Total economic value of marine biodiversity: Malaysia marine parks. https://wdpa.s3.amazonaws.com/Country_informations/MYS/TOTAL%20ECONOMIC%20VALUE%20OF%20MARINE%20BIODIVERSITY.pdf.

Mallow, E., & Fox, M. A. (2014). Phthalates and critically ill neonates: Device-related exposures and non-endocrine toxic risks. *Journal of Perinatology, 34*(12), 892–897.

Marcus, G. E. (1995). Ethnography in/of the world system: The emergence of multi-sited ethnography. *Annual Review of Anthropology, 24*, 95–117.

Markowitz, G. E. & Rosner, D. (2002). *Deceit and denial: The deadly politics of industrial pollution.* University of California Press.

Martin, C. B., Hales, C. M., Gu, Q., & Ogden, C. L. (2019). *Prescription drug use in the United States, 2015–2016* (NCHS Data Brief, No. 334). National Center for Health Statistics.

Mol, A. P. J., & Spaargaren, G. (2006). Toward a sociology of environmental flows: A new agenda for twenty-first-century environmental sociology In F. H. Buttel (Ed.), *Governing environmental flows: Global challenges to social theory* (pp. 39–82). MIT Press.

Mundt, K. A., Dell, L. D., Crawford, L., & Gallagher, A. E. (2017). Quantitative estimated exposure to vinyl chloride and risk of angiosarcoma of the liver and hepatocellular cancer in the US industry-wide vinyl chloride cohort: Mortality update through 2013. *Occupational and Environmental Medicine, 74*(10), 709–716.

Mustafa, M. (2011). The role of environmental impact assessment in addressing marine environmental issue arising from oil and gas activities: Examples from Malaysia. *International Proceedings of Chemical, Biological & Environmental Engineering (IPCBEE), 21*, 58–62.

National Cancer Institute Surveillance, Epidemiology and End Results Program. (2019). Cancer Stat Facts. https://seer.cancer.gov/statfacts/.

National Hospice and Palliative Care Organization. (2018). Facts & figures: Hospice care in America. https://www.nhpco.org/sites/default/files/public/Statistics_Research/2017_Facts_Figures.pdf.

National Hospice and Palliative Care Organization. (2020). Hospice facts & figures. https://www.nhpco.org/hospice-facts-figures/.

National Institute for Occupational Safety and Health. (2004). *Preventing occupational exposures to antineoplastic and other hazardous drugs in healthcare settings.* https://www.cdc.gov/niosh/docs/2004-165/pdfs/2004-165.pdf.

National Institute for Occupational Safety and Health. (2008). *Personal protective equipment for health care workers who work with hazardous drugs.* https://www.cdc.gov/niosh/docs/wp-solutions/2009-106/pdfs/2009-106.pdf.

National Institute for Occupational Safety and Health (2016). NIOSH list of antineoplastic and other hazardous drugs in healthcare settings, 2016. https://www.cdc.gov/niosh/docs/2016-161/.

National Institute for Occupational Safety and Health (2017). Health and safety practices survey of healthcare workers: Antineoplastic drugs—administration. https://www.cdc.gov/niosh/topics/healthcarehsps/antineoadmin.html.

Naveen, B., Mahapatra, D. M., Sitharam, T., Sivapullaiah, P., & Ramachandra, T. (2017). Physico-chemical and biological characterization of urban municipal landfill leachate. *Environmental Pollution, 220*, 1–12.

Nilsson, J., Blomberg, C., Holgersson, G., Carlsson, T., Bergqvist, M., & Bergström, S. (2017). End-of-life care: Where do cancer patients want to die? A systematic review. *Asia-Pacific Journal of Clinical Oncology, 13*(6), 356–364.

Ojemaye, C. Y., & Petrik, L. (2019). Pharmaceuticals in the marine environment: A review. *Environmental Reviews, 27*(2), 151–165.

Pacetti, P., Paganini, G., Orlandi, M., Mambrini, A., Pennucci, M. C., Del Freo, A., & Cantore, M. (2015). Chemotherapy in the last 30 days of life of advanced cancer patients. *Supportive Care in Cancer, 23*(11), 3277–3280.

Parajuli, J., Tark, A., Jao, Y.-L., & Hupcey, J. (2020). Barriers to palliative and hospice care utilization in older adults with cancer: A systematic review. *Journal of Geriatric Oncology, 11*(1), 8–16.

Patz, J. A., Campbell-Lendrum, D., Holloway, T., & Foley, J. A. (2005). Impact of regional climate change on human health. *Nature, 438*(7066), 310–317.

Pellow, D. N. (2002). *Garbage wars: The struggle for environmental justice in Chicago / David Naguib Pellow*. MIT Press.

Perreault, H. A., Semsar, K., & Godwin, J. (2003). Fluoxetine treatment decreases territorial aggression in a coral reef fish. *Physiology & Behavior, 79*(4–5), 719–724.

Pesatori, A. C., Consonni, D., Bachetti, S., Zocchetti, C., Bonzini, M., Baccarelli, A., & Bertazzi, P. A. (2003). Short- and long-term morbidity and mortality in the population exposed to dioxin after the "Seveso accident." *Industrial Health, 41*(3), 127–138.

Pierce, J., & Jameton, A. (2005). Response to Carrick. *Perspectives in Biology and Medicine, 48*(3), 458–463.

Pierce, J., & Kerby, C. (1999). The global ethics of latex gloves: Reflections on natural resource use in healthcare. *Cambridge Quarterly of Healthcare Ethics 8*(1), 98–107.

Pittet, D., Tarara, D., & Wenzel, R. P. (1994). Nosocomial bloodstream infection in critically ill patients: Excess length of stay, extra costs, and attributable mortality. *JAMA, The Journal of the American Medical Association, 271*(20), 1598.

Pope, T. M. (2018). Legal duties of clinicians when terminally ill patients with cancer or their surrogates insist on "futile" treatment. https://www.ascopost.com/issues/march-10-2018/legal-duties-of-clinicians-when-terminally-ill-patients-with-cancer-or-their-surrogates-insist-on-futile-treatment/.

Prochaska, J. D., Nolen, A. B., Kelley, H., Sexton, K., Linder, S. H., & Sullivan, J. (2014). Social determinants of health in environmental justice communities: Examining cumulative risk in terms of environmental exposures and social determinants of health. *Human and Ecological Risk Assessment: An International Journal, 20*(4), 980–994.

Risse, G. B. (1999). *Mending bodies, saving souls: A history of hospitals*. Oxford University Press.

Ristovski, Z. D., Miljevic, B., Surawski, N. C., Morawska, L., Fong, K. M., Goh, F., & Yang, I. A. (2012). Respiratory health effects of diesel particulate matter. *Respirology, 17*(2), 201–212.

Roser, M. (2017). Link between health spending and life expectancy: The US is an outlier. https://ourworldindata.org/the-link-between-life-expectancy-and-health-spending-us-focus.

Ruhoy, I. S., & Daughton, C. G. (2008). Beyond the medicine cabinet: An analysis of where and why medications accumulate. *Environment International, 34*(8), 1157–1169.

Sandhu, G. S., Frey, H. C., Bartelt-Hunt, S., & Jones, E. (2016). Real-world activity, fuel use, and emissions of diesel side-loader refuse trucks. *Atmospheric Environment, 129*, 98–104.
Santos-Burgoa, C., Matanoski, G. M., Zeger, S., & Schwartz, L. (1992). Lymphohematopoietic cancer in styrene-butadiene polymerization workers. *American Journal of Epidemiology, 136*(7), 843–854.
Sarma, S., Garcia-Garcia, G., Nandini, S., & Saucedo-Campos, A. (2017). Effects of anti-diabetic pharmaceuticals to non-target species in freshwater ecosystems: A review. *Journal of Environmental Biology, 38*(6), 1249–1254.
Sathishkumar, P., Meena, R. A. A., Palanisami, T., Ashokkumar, V., Palvannan, T., & Gu, F. L. (2020). Occurrence, interactive effects and ecological risk of diclofenac in environmental compartments and biota-a review. *Science of the Total Environment, 698*, 134057.
Saunders, D. C. M., Summers, D. H., & Teller, N. (1981). *Hospice: The living idea*. WB Saunders.
Savage, J. (2000). Ethnography and health care. *BMJ (Clinical Research Ed.), 321*(7273), 1400.
Schnaiberg, A. (1980). *The environment: From surplus to scarcity*. Oxford University Press.
Scott, R. D. (2009). The direct medical costs of healthcare-associated infections in US hospitals and the benefits of prevention. National Center for Preparedness, Detection, and Control of Infectious Diseases (U.S.). Division of Healthcare Quality Promotion. https://stacks.cdc.gov/view/cdc/11550.
Sen, T., & Samanta, S. K. (2014). Medicinal plants, human health and biodiversity: A broad review. *Biotechnological Applications of Biodiversity. 147*, 59–110. https://doi.org/10.1007/10_2014_273.
Skokan, E. (2021, June). *Glove story: Global glove production amidst the COVID-19 pandemic*. U.S. International Trade Commission, Executive Briefing on Trade. https://www.usitc.gov/publications/332/executive_briefings/ebot_glove_story_global_glove_production_amidst_covid-19_pandemic.pdf.
Smith, S., Brick, A., O'Hara, S., & Normand, C. (2014). Evidence on the cost and cost-effectiveness of palliative care: A literature review. *Palliative Medicine, 28*(2), 130–150. https://doi.org/10.1177/0269216313493466
Starr, P. (1982). *The social transformation of American medicine*. Basic Books.
Sudnow, D. (1967). *Passing on: The social organization of dying*. Prentice-Hall.
Sydbom, A., Blomberg, A., Parnia, S., Stenfors, N., Sandström, T., & Dahlén, S. E. (2001). Health effects of diesel exhaust emissions. *European Respiratory Journal, 17*(4), 733–746.
Tait, P. W., Brew, J., Che, A., Costanzo, A., Danyluk, A., Davis, M., Khalaf, A., McMahon, K., Watson, A., Rowcliff, K., & Bowles, D. (2020). The health impacts of waste incineration: A systematic review. *Australian and New Zealand Journal of Public Health, 44*(1), 40–48.
Tan, K., Johnson, B., Goh, B., Tun, K., Low, J., Gin, K., Ting, Y. P., Obbard, J., Tan, H. M., Mathew, M., & Chou, L.M. (1999). An assessment of the impact of the *Evoikos* oil spill on the marine environment in Singapore. *Singapore Maritime and Port Journal*, 69–81.
Tchobanoglous, G., & Kreith, F. (2002). *Handbook of solid waste management*. McGraw-Hill.

Temel, J. S., Greer, J. A., Muzikansky, A., Gallagher, E. R., Admane, S., Jackson, V. A., Dahlin, C. M., Blinderman, C. D., Jacobsen, J., Pirl, W. F., Billings, J. A., & Lynch, T. J. (2010). Early palliative care for patients with metastatic non-small-cell lung cancer. *New England Journal of Medicine, 363*(8), 733–742.

Tennison, I., Roschnik, S., Ashby, B., Boyd, R., Hamilton, I., Oreszczyn, T., Owen, A., Romanello, M., Ruyssevelt, P., Sherman, J. D., & Smith, A. Z. (2021). Health care's response to climate change: A carbon footprint assessment of the NHS in England. *Lancet Planetary Health, 5*(2), e84–e92.

Teno, J. M., Clarridge, B. R., Casey, V., Welch, L. C., Wetle, T., Shield, R., & Mor, V. (2004). Family perspectives on end-of-life care at the last place of care. *JAMA, The Journal of the American Medical Association, 291*(1), 88.

Tijani, J. O., Fatoba, O. O., Babajide, O. O., & Petrik, L. F. (2016). Pharmaceuticals, endocrine disruptors, personal care products, nanomaterials and perfluorinated pollutants: A review. *Environmental chemistry letters, 14*(1), 27–49.

Tikkanen, R., & Abrams, M. K. (2020). *U.S. health care from a global perspective, 2019: Higher spending, worse outcomes?* https://www.commonwealthfund.org/publications/issue-briefs/2020/jan/us-health-care-global-perspective-2019.

Tseng, Y. D., Gouwens, N. W., Lo, S. S., Halasz, L. M., Spady, P., Mezheritsky, I., & Loggers, E. (2018). Use of radiation therapy within the last year of life among cancer patients. *International Journal of Radiation Oncology, Biology, Physics, 101*(1), 21–29.

Tulsky, J. A. (2015). Improving quality of care for serious illness: Findings and recommendations of the Institute of Medicine report on dying in America. *JAMA Internal Medicine, 175*(5), 840–841.

United Church of Christ Commission for Racial Justice. (1987). *Toxic wastes and race in the United States: A national report on the racial and socio-economic characteristics of communities with hazardous waste sites.* Commission for Racial Justice, United Church of Christ, New York. https://www.nrc.gov/docs/ML1310/ML13109A339.pdf.

U.S. Department of Justice Drug Enforcement Administration. (2020). Drug disposal information. https://www.deadiversion.usdoj.gov/drug_disposal/index.html.

U.S. Department of Labor. (2019). Minimum wage laws in the states. https://www.dol.gov/whd/minwage/america.htm.

U.S. Department of Labor, Bureau of Labor Statistics. (2019). Labor force statistics from the current population survey. https://www.bls.gov/cps/cpsaat18.htm.

U.S. Food and Drug Administration. (2020). Where and how to dispose of unused medicines. https://www.fda.gov/consumers/consumer-updates/where-and-how-dispose-unused-medicines.

U.S. Global Change Research Program. (2016). The impacts of climate change on human health in the United States: A scientific assessment. http://dx.doi.org/10.7930/J0R49NQX.

U.S. Senate Committee on Finance (2019). Grassley urges HHS, FDA to implement unannounced inspections of foreign drug manufacturing facilities. https://www.finance.senate.gov/chairmans-news/grassley-urges-hhs-fda-to-implement-unannounced-inspections-of-foreign-drug-manufacturing-facilities.

Urry, J. (2000). *Sociology beyond societies: Mobilities for the twenty-first century.* Routledge.

Urry, J. (2002). Mobility and proximity. *Sociology, 36*(2), 255–274.

Urry, J. (2003). *Global complexity*. Polity.
Valenti, T. W., Jr., Gould, G. G., Berninger, J. P., Connors, K. A., Keele, N. B., Prosser, K. N., & Brooks, B. W. (2012). Human therapeutic plasma levels of the selective serotonin reuptake inhibitor (SSRI) sertraline decrease serotonin reuptake transporter binding and shelter-seeking behavior in adult male fathead minnows. *Environmental Science & Technology, 46*(4), 2427–2435.
Van Vliet, P., Knape, M., de Hartog, J., Janssen, N., Harssema, H., & Brunekreef, B. (1997). Motor vehicle exhaust and chronic respiratory symptoms in children living near freeways. *Environmental Research, 74*(2), 122–132.
Vatovec, C., Kolodinsky, J., Callas, P., Hart, C., & Gallagher, K. (2021). Pharmaceutical pollution sources and solutions: Survey of human and veterinary medication purchasing, use, and disposal. *Journal of Environmental Management, 285*, 112106.
Vatovec, C., Van Wagoner, E., & Evans, C. (2017). Investigating sources of pharmaceutical pollution: Survey of over-the-counter and prescription medication purchasing, use, and disposal practices among university students. *Journal of Environmental Management, 198*, 348–352.
Villanueva, A., & Wenzel, H. (2007). Paper waste—recycling, incineration or landfilling? A review of existing life cycle assessments. *Waste Management, 27*(8), S29–S46.
Vogel, M. J. (1980). *The invention of the modern hospital, Boston, 1870–1930*. University of Chicago Press.
Vrijheid, M. (2000). Health effects of residence near hazardous waste landfill sites: A review of epidemiologic literature. *Environmental Health Perspectives, 108 Suppl 1*, 101.
Vrijheid, M. (2009). Landfill sites and congenital anomalies—have we moved forward? *Occupational and Environmental Medicine, 66*(2), 71. https://doi.org/10.1136/oem.2008.040998
Walsh, V., & Goodman, J. (1999). Cancer chemotherapy, biodiversity, public and private property: The case of the anti-cancer drug Taxol. *Social Science & Medicine, 49*(9), 1215–1225.
Weaver, B. A. (2014). How Taxol/paclitaxel kills cancer cells. *Molecular Biology of the Cell, 25*(18), 2677–2681.
Wernet, G., Conradt, S., Isenring, H. P., Jiménez-González, C., & Hungerbühler, K. (2010). Life cycle assessment of fine chemical production: A case study of pharmaceutical synthesis. *International Journal of Life Cycle Assessment, 15*(3), 294–303.
Wise, J. (2021). Climate crisis: Over 200 health journals urge world leaders to tackle "catastrophic harm." *BMJ: British Medical Journal (Online), 374*. https://doi.org/10.1136/bmj.n217.
World Health Organization. (2000). Guidelines for healthcare equipment donations. https://www.who.int/hac/techguidance/pht/1_equipment%20donationbuletin82WHO.pdf.
World Health Organization. (2018). Ambient (outdoor) air pollution. https://www.who.int/news-room/fact-sheets/detail/ambient-(outdoor)-air-quality-and-health.
Wright, A. A., Keating, N. L., Ayanian, J. Z., Chrischilles, E. A., Kahn, K. L., Ritchie, C. S., Weeks, J. C., Earle, C. C., & Landrum, M. B. (2016). Family perspectives on

aggressive cancer care near the end of life. *JAMA, 315*(3), 284–292. http://jama.jamanetwork.com/data/Journals/JAMA/934869/joi150166.pdf.

Zainab, S. M., Junaid, M., Xu, N., & Malik, R. N. (2020). Antibiotics and antibiotic resistant genes (ARGs) in groundwater: A global review on dissemination, sources, interactions, environmental and human health risks. *Water Research, 187*, 116455.

Zimring, C. A. (2017). *Clean and white: A history of environmental racism in the United States*. NYU Press.

Index

acid rain, 82
acrylonitrile, 62–64
acute myeloid leukemia, 19
acyclovir, 77–78, 137
African Americans, 33, 89–90, 143n9, 147n15
aggressive care, 15, 18–20, 93, 101, 103, 107–108, 113, 141n9
agricultural land degradation, 63
air pollution, 32, 40, 82, 89
allopathic. *See* conventional care
American Hospitals Association, 4
American Nurses Association, 4
American Society of Clinical Oncology, 19, 103
amphipods, 95
angiosarcoma of the liver, 69
antianxiety medication, 95, 148n21
antibiotic, 78–79, 137
antidepressant, 95, 148n21
antineoplastic drugs, 84, 86, 146n8, 146n9, 146n10, 146n11. *See also* chemotherapy
antiviral, 77, 79, 137
Approaching Death Institute of Medicine report, 16
aquatic ecosystems and species, 2, 68, 92, 95, 118, 133, 147n20
aquatic pollution, 8, 91
armpit of the hospital, 34–35
automation of medical supply reprocessing, 75
automotive industry, 63, 81, 145n4
autonomy. *See* patient autonomy

bereavement, 101, 112, 114, 148n2
best practices, 5, 29
bioaccumulation, 68, 95
biodiversity, 2, 57–58, 64, 79
biohazard, 36, 39
bioprospecting, 80
bladder cancer, 64
blue bin recycling, 46
body fluids, 39, 66
bone marrow transplant, 18, 55, 66, 83, 101, 137, 142n4
boundaries. *See* planetary boundaries
brain cancer, 18–19, 104
breast cancer, 18–19, 80, 83
butadiene, 62–64

Cadillac of Dumps, 89
Canada, 60, 139n4
cancer, 1
Cancer Alley, 69, 145n9
carbon dioxide, 31, 143n6, 145n4
carbon emissions. *See* emissions
carbon sequestration, 64
cardiovascular illness, 69, 82

"Care without Carbon: The Road to Sustainability in U.S. Health Care," 5
cargo ships, 2, 65
Carrick, Paul, 6–7, 140n10
chaplain, 25, 111–112
chemical policy, 145n12
chemotherapy: biological effects of, 83–84; cancer treatment, 18–19, 55, 78–80, 101–102, 148n5, 141n9, 107–108; environmental consequences, 13; occupational exposures, 8, 62, 83–87, 117, 146n9, 146n10; personal protective equipment, 83, 85–86; pre-priming intravenous bags, 86–87. *See also* clinical trial
China, 60, 62, 82, 146n5, 146n6
clean area, 49, 52, 53, 114, 115
climate change, 2–3, 31, 63–64, 120, 139n2, 142n5
clinical practices, 1
clinical trial, 18, 80, 99, 108
coal, 82, 146n5
colon cancer, 18–19, 69, 87
communication. *See* death and dying
communities of color, 31–32
compactor, 35, 41
comparative analysis, 7, 14, 123
Congress, 16
constipation, 79, 83
conventional care: goals of care, 14; introduction to research site, 14–19
coral reef, 64
Costa Rica, 60
cost of care. *See* healthcare expense
cost savings, 140n9
co-trimaxazole, 77
COVID-19 pandemic, 64–65, 139n2
CPR, 102
cradle to grave, 57–58, 60–61, 65, 70, 127–128, 131
cumulative impacts, 7–8, 17, 24, 65, 88, 101, 133, 139n1

death and dying, open dialog between patients and clinicians, 8, 15, 100, 103, 106–107, 113–114. *See also* philosophies of care
decontamination, 33–38
DEHP, 59, 68, 70, 145n10, 145n11
Denmark, 60
design. *See* facility design
diabetes, 69
diagnosis, 15, 19, 23, 93, 101, 107
die at home. *See* location of death
dignity, 7, 100
dioxin, 4, 41, 68–70, 89
dip-line, 62–63, 65
discharge cleaning, 27
disease progression, 87, 105, 112, 141n13
disinfection, 1, 17–18, 27–28, 35, 72
division of labor, 52, 53
donation. *See* medical supplies
dressing kit, 55–56, 142n4
drilling, of oil, 63–64
drinking water, 2, 58, 94, 147n19
dripable, pourable, squeezable, 39, 44
drug disposal, 114
drug diversion, 8, 88, 90–92, 95–96, 118; policy, 87, 91–92
Drug Enforcement Agency. *See* U.S. Department of Justice Drug Enforcement Agency
drug take-back program, 96, 118

ecological boundaries. *See* planetary boundaries
ecological footprint, 1, 2, 4–5, 13, 60, 131, 139n5
economic, 8, 18, 20, 29, 101, 114, 116, 142n1
education campaign. *See* training for waste disposal
electricity, 2, 31, 143n18
electric vehicles, 115
elevator, 12, 17, 34, 37–38
Emelle, Alabama, 89–90, 147n16
emergency: emergency room, 30, 101, 141n9; no emergency in hospice, 22, 52, 76
emissions, 2–3, 29, 31–32, 40–41, 82, 139n2, 142n5, 143n6, 145n4, 146n5
endangered species, 81
end-of-life care, 1, 7–8, 11–16, 23, 100–103
energy, 1–5, 13, 29, 140n9, 142n3; in pharmaceutical manufacturing, 81–82, 145n4
Energy Star program, 82, 145n4

environmental boundaries. *See* planetary boundaries
environmental flows, 6, 58, 127–132, 134, 149n1, 149n2
environmental justice, 70, 90, 147n14, 147n16
environmentally preferred supply purchasing, 4, 5, 13, 72, 116
Environmental Protection Agency (EPA), 4, 31, 41, 81, 96, 118, 142n1, 141n3, 143n6, 143n11, 143n14, 145n4, 146n6, 146n12, 148n22
Environmental Services Department, 33–34, 43, 130
environmental services staff, 11, 17, 33–34, 37, 51–52, 88, 115, 130
ethical: conflict between medical care and community health, 2, 6, 57, 140n10; debate regarding Taxol, 80; implications of medical supply donation, 51, 53, 143n18; life cycle analysis, 8, 57, 61
ethnography, 1, 123–125
euphemisms, 105
externalities, 3
extraction, 2, 57, 63, 68

facility design, 5, 20, 53
factories, 2, 6, 62
fisheries, 64
flourishing environment, 4
flows. *See* environmental flows
"flush list" for medication disposal, 96, 148n22
Food and Drug Administration, 73, 96, 145n10, 148n22
food chain, 68, 95
food insecurity, 58
food supply contamination, 69
futile treatments, 19, 113, 141n9

gabapentin, 78, 137
germ theory of disease, 15
goals of care, 14, 16, 18, 20–24, 79, 99, 104–106, 113, 133, 140n5, 141n9
green cleaning chemicals, 4–5
Greenhealth Academy, 4
Green Healthcare Institutions: Health, Environment, and Economics, 5
greenhouse gas emissions. *See* emissions
greening. *See* healthcare greening
green pharmacy, 118
grief counseling, 25
gross domestic product, 3, 140n3
group purchasing organization, 5, 72, 116
gurney, 12, 112

habitat fragmentation, 63
harvesting of materials, 2, 29, 64, 79–81, 145n2
hazardous waste, 42, 58, 88–90, 117, 146n12, 147n13. *See also* incineration
healthcare expense, 3, 5, 13, 16, 18, 20, 23, 52, 140n3, 141n11
healthcare greening, 4, 6, 140n9
Health Care Without Harm, 4, 139n5, 145n10
health effects: of chemotherapy, 84; of climate change, 3; of hazardous waste incineration, 147n13; of material extraction, 65; of toxic exposures in medical supply chains, 2, 68–69, 146n5
Health Insurance Portability and Accountability Act, 29, 131
health outcomes, 3, 24, 29, 32, 49, 63, 84, 132, 139n2, 146n12
health spending per capita, 3
hematology, 18–19, 77
hematoma, 99–100
Hevea tree, 57
history: of end-of-life care, 7, 14, 16, 140n6; of healthcare greening, 4; of racism in waste management, 143n9, 147n14
HIV, 36
Hodgkin's disease, 19
Hoffman-La Roche, 69
hospice: barriers to entry, 100; benefits to patients and families, 24, 100; goals of care, 23; history of, 15–16; introduction to research site, 22–23; philosophy of, 14; transitioning when medically appropriate, 9, 100–103, 109, 113–114, 118
hospital-acquired infection. *See* nosocomial infection
human rights, 65
hydrogen chloride, 41
hydromorphone, 87, 137

illicit drugs, 90. *See also* drug diversion
immigrants, 33
immune compromised, 79, 83
immunosuppressant, 77, 137
IMS Institute for Healthcare Informatics, 145n1
incineration, 4, 29, 40–41, 68, 88–90, 96, 117, 141n1, 142n1, 146n12; facilities, 41, 58, 89
Indonesia, 57, 62
industrialization, 15
infection control, 48–50, 53, 65–66, 71, 114–115, 131–133, 142, 143; cost of treatment, 71, 143n15, 143n16; environmental flows of, 131–133; and glove use, 65; policy, 27; and product evaluation, 71; standards, 28, 48, 53, 66, 115
infectious waste, 29–31, 35–41, 52, 58, 89, 130
infrastructure, 3, 114
Institute of Medicine, 5, 16, 141n10
institutional governance, 8, 67, 70, 133
intensive care unit (ICU), 20, 21, 70, 101, 101, 105–106, 113, 149n2. *See also* location of death
International Agency for Research on Cancer, 64, 68
International Trade Commission, 64
International Union for Conservation of Nature, 81
intravenous pump, 2, 60, 67, 73
intravenous tubing, 2, 136
Iran, 63
Iraq, 63
Israel, 60

Jameton, Andrew, 6, 139n1, 140n10
Japan, 3, 60
Joint Commission, 97
justice. *See* environmental justice; social justice

Kerby, Christina, 8, 57–58, 61, 65
Kubler-Ross, Elizabeth, 16, 140n2

latex, 6, 57–59, 61–63, 65–66
Latinx, 89, 147n15
laundry, 1, 28, 45, 72
length of stay, 21, 24–25, 101–102, 133, 141n14, 142n4
life-cycle analysis, 8, 57, 82, 141n1
life expectancy, 3, 139n4
life-limiting illness, 14, 16, 104, 133
life-prolonging therapy, 9, 16, 21, 25, 100, 103, 105, 107
linens, 28, 86. *See also* laundry
loading dock, 35, 38, 58, 60, 65, 90, 130–131
location of death, 24, 100, 106, 113, 141n13, 149n2
logging, 80
low income, 32, 58, 89
lung cancer, 18–19, 64, 69, 80, 89, 111

"made in the U.S.A.," 60, 69
Malaysia, 57, 60, 62–65
mangrove forests, 64
marginalized communities, 33
Materials Recovery Facility (MRF), 46
mechanical sorting. *See* recycling
mechanical ventilation, 16, 102
Medicaid, 49, 140n3, 141n12
medical decision making, 1–2, 5–7, 12–13, 112, 114, 125, 127, 140n10
medicalization of dying, 15–16
medically appropriate care, 9, 100–102, 113–114, 118, 133
medical supplies: donation, 48, 50–53, 114; economic cost, 56, 61–62, 65, 71, 74–75, 116, 131. *See also* environmentally preferred supply purchasing
Medical Waste Tracking Act of 1988, 36
Medicare, 13, 16, 52, 102–103, 108, 140n3, 141n12; reimbursement, 49, 71
mercury, 4, 41, 58, 68, 70
metal, 1, 12, 41, 46, 58, 68, 73, 89, 142n2
methane, 29, 31, 143n6
methicillin-resistant *Staphylococcus aureus,* 67
Mexico, 60
microwave, 41
military-industrial complex, 63, 129
"mobile meds" device, 77–78, 87
money, 3, 49, 56, 72. *See also* political economy
morgue, 12, 112
morphine, 88, 91–92, 137
Morrisonville, Louisiana, 69
mortality, 3, 15, 101

MRSA. *See* methicillin-resistant *Staphylococcus aureus*
municipal trash, 29–32, 34–36, 38–40, 44

narcotics, 87–88, 90–93, 96–97, 117. *See also* drug diversion
National Academies of Sciences Institute of Medicine. *See* Institute of Medicine
National Cancer Institute, 80, 141n8
National Climate and Health Assessment, 3
National Institute of Occupational Safety and Health, 97, 146n8, 146n10, 146n11
natural resources, 1, 13, 24, 128
nature of dying, 15. *See also* bioprospecting
nausea, 79, 83, 107
needle, 36, 39, 47, 75
neuropathic pain, 78, 137
new immigrants. *See* immigrants
nitrile gloves, 8, 28, 56, 59; forced labor in production, 65; gold standard for chemotherapy delivery, 62, 83, 85; and latex allergies, 66; life cycle of, 61–65
non-Hodgkin's lymphoma, 19, 89
non-infectious waste, 45
nonrecyclable, 134
nosocomial infection, 15, 49, 71; cost of care, 49
"not in my backyard" (NIMBY), 32
nursing staff: nursing assistant, 17, 25, 37, 48; staff meetings, 44–45, 124, 134

Occupational Safety and Health Administration, 97
old-growth forests, 80–81
oncology, 18–20, 84, 87, 99, 103, 107, 142n4
On Death and Dying, 16, 140n2
open communication. *See* death and dying
operating room, 39, 43–44, 46, 72
Organization for Economic Cooperation and Development, 3
outpatient, 18–19, 84, 88
ovarian cancer, 80
overdispensing, 8, 117
overprescribing, 8, 92, 117, 147n18
over-the-counter medications, 78, 92
OxyContin, 93

Pacific yew, 80–81
pain. *See* symptom management
palliative care: consultation service, 20–21, 104–107, 133; goals of care, 14, 16; introduction to research site, 20–21
pancreatic cancer, 19
paper recycling, 29
Paris Climate Agreement, 82
particulate matter, 31, 41, 58, 82, 146n6
patient autonomy, 6–7, 100, 140n10
patient census, 17
patient day, 30, 136–137
patient preferences, 9, 24, 100–101, 113–114, 141n7, 141n13, 148n1, 149n2
patient satisfaction, 5, 49, 100, 107, 140n9
peripherally inserted central catheter (PICC), 55–56
personal protective equipment, 42, 84, 87, 117
petroleum, 129; in nitrile production, 63; in PVC production, 67
pharmaceutical, as hazardous waste, 88–89. *See also* drug disposal; drug diversion; overdispensing; overprescribing
pharmaceutical waste: detection in surface waters, 94; environmental fate of, 94
philosophies of care, 23, 31, 79, 94, 100, 112, 140n5
photochemical smog, 82
Pierce, Jessica, 6, 8, 57–58, 61, 65, 139n1, 140n10
planetary boundaries, 3–4, 9
planetary health, 6, 101, 140n10
plan of care, 104
points of intervention, 1, 8, 30, 114, 116, 127–128, 130, 132, 149n2
policy: drug disposal directives, 96, 114, 148n22; institutional as basis for palliative care, 133. *See also* chemical policy; infection control
political economy, 6, 38, 74, 128
polyvinyl chloride, 61, 67–70, 145n11
Port Arthur, Texas, 89–90, 147n15
posaconazole sulfate, 78
Practice Greenhealth, 4, 139n6
prednisone, 77
prepackaged dressing kit. *See* dressing kit

prescription medication, 78–79, 93, 118, 145n1
procedural kit, 116. *See also* dressing kit
procession, 111–112
procurement, 5, 57, 116
product evaluation, 70–71
profitability, 62, 101, 148n3
prognosis, 23–24, 93, 99–100, 103–104, 149n6
prophylactic medication, 79
Prozac, 95, 148n21
pseudopersistence, 92
purchasing agent, 47, 71
PVC. *See* polyvinyl chloride

racism, 89–90, 143n9. *See also* environmental justice; social justice
ranitidine, 77
recyclable, 40, 43–47, 87
recycling, 5, 29–31, 33, 35, 42–48, 52–53, 114–115, 125, 134, 141n1, 142n2; administrative support, 45; contamination of, 44; cost, 45; mechanical sorting, 47, 115; program failure, 46; program success, 45; staff support, 44; starting a program, 43; time required, 43–44; troubleshooting problems, 44
red bag, 36, 39–40, 42, 44. *See also* infectious waste
registered nurse, 17
regulation: environmental rules abroad, 63, 81–82; and occupational health, 58, 69; of recycling, 40; and waste reduction, 40. *See also* drug diversion; infection control
rehabilitation, 99
remission, 112
reprocessing, 51, 72–75, 116
respiratory illness: and climate change, 3; and diesel emissions, 142n5; and dioxins, 69; and pharmaceutical manufacturing, 82; resulting from occupational exposure to acrylonitrile, 64
reusable medical supplies, 51, 59, 72–74, 116
Reveilletown, Louisiana, 69
Roundtable on Environmental Health Sciences, Research, and Medicine, 5
Russia, 63
satisfaction. *See* patient satisfaction
Saudi Arabia, 63
Saunders, Dame Cicely, 15–16, 140n5
Schedule II controlled substances, 87, 137
Seveso, Italy, 69
sharps, 39–40, 42, 115. *See also* needle
Singapore, 60
skilled nursing facility, 21, 99–100
social connection, 38, 53, 115
social disconnection, 37
social justice, 70
social workers, 17, 20, 25, 105, 111, 124
soiled utility room, 28, 34, 37–39
spotted owl, 80
staff meeting. *See* nursing staff
stainless steel, 41
stapler, 41
statins, 93
St. Christopher's in the Field (hospice), 15–16
sterile/sterility, 43, 56, 71
sterilize/sterilizer, 41, 56, 73–74
steroid, 77, 137
Stockholm Convention on Persistent Organic Pollutants, 68
suction tubing, 35
SUPPORT study, 16
surgery, 15, 19, 30–31, 41, 71, 73
survivorship, 95, 112, 148n21
sustainability, 1
sustainable procurement, 116
swamp, 35
sweatshop, 66
Sweden, 60, 139n4
Switzerland, 3, 139n4
symptom management, 14, 16, 19–20, 23, 104
synthetic rubber, 62, 64

Taxol, 79–81
Taxus brevifolia, 80
Taxus contorta, 81
technological advancement, 2, 15, 115
terminal illness, 1, 7, 14–16, 24–25, 93, 112–113, 118, 123, 141n10
testicular, 104
Thailand, 57, 60
timber, 80
tourism, 64

toxin, 2, 4, 58
training for waste disposal, 33, 35, 37–38, 40, 42–48, 52–53
transplant, 35, 71. *See also* bone marrow transplant
trash compactor. *See* compactor
trust, as necessary in transitioning care, 103–104, 109
tubing. *See* intravenous tubing
tyvec, 44

unintended consequences, 1, 8, 11, 13, 29, 58, 61, 64; comparison of medical care options, 113, 123; of medical supply chains, 75; of plastics, 67; related to environmental flows, 133; of waste disposal, 115–116
unnecessary treatment, 23, 133
U.S. Customs and Border Protection, 65
U.S. Department of Agriculture, 80
U.S. Department of Justice Drug Enforcement Agency, 96, 148n22
U.S. Geological Survey, 94, 147n19

vaccine, 65
vinyl: chloride, 69–70, 145n9; gloves, 56

Wald, Florence, 16
waste: cost, 32, 40, 41, 45, 50, 52, 53, 133; disposal, 2, 31, 37, 39, 45, 52, 115, 118, 132; municipal, 130; of pharmaceuticals, 88; reduction, 40, 52, 56, 114, 115, 134, 140n9, 143n17; stream, 8, 29–31, 33–37, 39–41, 44, 47, 52, 58. *See also* training for waste disposal
wastewater, pharmaceutical pollution in, 91, 94, 95–96, 147n20
water: contamination from landfill leachate, 58; contamination in latex production, 57–58, 65; mercury pollution in plastic manufacturing, 68–69; pharmaceutical pollution in, 147; pollution in nitrile production, 63; used in medical supply sterilization, 74; used in paper recycling, 29
widower effect, 101
Workshop on Environmental Sustainability in Clinical Care, 5
World Health Organization, 64, 68, 82, 143n19, 146n6

Yale University's Center on Climate Change and Health, 5

About the Author

CHRISTINE VATOVEC is a research assistant professor at the University of Vermont Larner College of Medicine, a fellow at the Gund Institute for Environment, and an award-winning lecturer at the Rubenstein School of Environment and Natural Resources.

Available titles in the Critical Issues in Health and Medicine series:

Emily K. Abel, *Prelude to Hospice: Florence Wald, Dying People, and Their Families*
Emily K. Abel, *Suffering in the Land of Sunshine: A Los Angeles Illness Narrative*
Emily K. Abel, *Tuberculosis and the Politics of Exclusion: A History of Public Health and Migration to Los Angeles*
Marilyn Aguirre-Molina, Luisa N. Borrell, and William Vega, eds., *Health Issues in Latino Males: A Social and Structural Approach*
Anne-Emanuelle Birn and Theodore M. Brown, eds., *Comrades in Health: U.S. Health Internationalists, Abroad and at Home*
Karen Buhler-Wilkerson, *False Dawn: The Rise and Decline of Public Health Nursing*
Susan M. Chambré, *Fighting for Our Lives: New York's AIDS Community and the Politics of Disease*
Stephen M. Cherry, *Importing Care, Faithful Service: Filipino and Indian American Nurses at a Veterans Hospital*
Brittany Clair, *Carrying On: Another School of Thought on Pregnancy and Health*
James Colgrove, Gerald Markowitz, and David Rosner, eds., *The Contested Boundaries of American Public Health*
Elena Conis, Sandra Eder, and Aimee Medeiros, eds., *Pink and Blue: Gender, Culture, and the Health of Children*
Cynthia A. Connolly, *Children and Drug Safety: Balancing Risk and Protection in Twentieth-Century America*
Cynthia A. Connolly, *Saving Sickly Children: The Tuberculosis Preventorium in American Life, 1909–1970*
Brittany Cowgill, *Rest Uneasy: Sudden Infant Death Syndrome in Twentieth-Century America*
Patricia D'Antonio, *Nursing with a Message: Public Health Demonstration Projects in New York City*
Kerry Michael Dobransky, *Managing Madness in the Community: The Challenge of Contemporary Mental Health Care*
Tasha N. Dubriwny, *The Vulnerable Empowered Woman: Feminism, Postfeminism, and Women's Health*
Edward J. Eckenfels, *Doctors Serving People: Restoring Humanism to Medicine through Student Community Service*
Julie Fairman, *Making Room in the Clinic: Nurse Practitioners and the Evolution of Modern Health Care*
Jill A. Fisher, *Medical Research for Hire: The Political Economy of Pharmaceutical Clinical Trials*
Charlene Galarneau, *Communities of Health Care Justice*
Alyshia Gálvez, *Patient Citizens, Immigrant Mothers: Mexican Women, Public Prenatal Care and the Birth Weight Paradox*
Laura E. Gómez and Nancy López, eds., *Mapping "Race": Critical Approaches to Health Disparities Research*

Janet Greenlees, *When the Air Became Important: A Social History of the New England and Lancashire Textile Industries*
Gerald N. Grob and Howard H. Goldman, *The Dilemma of Federal Mental Health Policy: Radical Reform or Incremental Change?*
Gerald N. Grob and Allan V. Horwitz, *Diagnosis, Therapy, and Evidence: Conundrums in Modern American Medicine*
Rachel Grob, *Testing Baby: The Transformation of Newborn Screening, Parenting, and Policymaking*
Mark A. Hall and Sara Rosenbaum, eds., *The Health Care "Safety Net" in a Post-Reform World*
Laura L. Heinemann, *Transplanting Care: Shifting Commitments in Health and Care in the United States*
Rebecca J. Hester, *Embodied Politics: Indigenous Migrant Activism, Cultural Competency, and Health Promotion in California*
Laura D. Hirshbein, *American Melancholy: Constructions of Depression in the Twentieth Century*
Laura D. Hirshbein, *Smoking Privileges: Psychiatry, the Mentally Ill, and the Tobacco Industry in America*
Timothy Hoff, *Practice under Pressure: Primary Care Physicians and Their Medicine in the Twenty-First Century*
Beatrix Hoffman, Nancy Tomes, Rachel N. Grob, and Mark Schlesinger, eds., *Patients as Policy Actors*
Ruth Horowitz, *Deciding the Public Interest: Medical Licensing and Discipline*
Powel Kazanjian, *Frederick Novy and the Development of Bacteriology in American Medicine*
Claas Kirchhelle, *Pyrrhic Progress: The History of Antibiotics in Anglo-American Food Production*
Rebecca M. Kluchin, *Fit to Be Tied: Sterilization and Reproductive Rights in America, 1950–1980*
Jennifer Lisa Koslow, *Cultivating Health: Los Angeles Women and Public Health Reform*
Jennifer Lisa Koslow, *Exhibiting Health: Public Health Displays in the Progressive Era*
Susan C. Lawrence, *Privacy and the Past: Research, Law, Archives, Ethics*
Bonnie Lefkowitz, *Community Health Centers: A Movement and the People Who Made It Happen*
Ellen Leopold, *Under the Radar: Cancer and the Cold War*
Barbara L. Ley, *From Pink to Green: Disease Prevention and the Environmental Breast Cancer Movement*
Sonja Mackenzie, *Structural Intimacies: Sexual Stories in the Black AIDS Epidemic*
Stephen E. Mawdsley, *Selling Science: Polio and the Promise of Gamma Globulin*
Frank M. McClellan, *Healthcare and Human Dignity: Law Matters*

Michelle McClellan, *Lady Lushes: Gender, Alcohol, and Medicine in Modern America*
David Mechanic, *The Truth about Health Care: Why Reform Is Not Working in America*
Richard A. Meckel, *Classrooms and Clinics: Urban Schools and the Protection and Promotion of Child Health, 1870–1930*
Terry Mizrahi, *From Residency to Retirement: Physicians' Careers over a Professional Lifetime*
Manon Parry, *Broadcasting Birth Control: Mass Media and Family Planning*
Alyssa Picard, *Making the American Mouth: Dentists and Public Health in the Twentieth Century*
Heather Munro Prescott, *The Morning After: A History of Emergency Contraception in the United States*
Sarah B. Rodriguez, *The Love Surgeon: A Story of Trust, Harm, and the Limits of Medical Regulation*
David J. Rothman and David Blumenthal, eds., *Medical Professionalism in the New Information Age*
Andrew R. Ruis, *Eating to Learn, Learning to Eat: School Lunches and Nutrition Policy in the United States*
James A. Schafer Jr., *The Business of Private Medical Practice: Doctors, Specialization, and Urban Change in Philadelphia, 1900–1940*
Johanna Schoen, ed., *Abortion Care as Moral Work: Ethical Considerations of Maternal and Fetal Bodies*
David G. Schuster, *Neurasthenic Nation: America's Search for Health, Happiness, and Comfort, 1869–1920*
Karen Seccombe and Kim A. Hoffman, *Just Don't Get Sick: Access to Health Care in the Aftermath of Welfare Reform*
Leo B. Slater, *War and Disease: Biomedical Research on Malaria in the Twentieth Century*
Piper Sledge, *Bodies Unbound: Gender-Specific Cancer and Biolegitimacy*
Dena T. Smith, *Medicine over Mind: Mental Health Practice in the Biomedical Era*
Kylie M. Smith, *Talking Therapy: Knowledge and Power in American Psychiatric Nursing*
Matthew Smith, *An Alternative History of Hyperactivity: Food Additives and the Feingold Diet*
Paige Hall Smith, Bernice L. Hausman, and Miriam Labbok, *Beyond Health, Beyond Choice: Breastfeeding Constraints and Realities*
Susan L. Smith, *Toxic Exposures: Mustard Gas and the Health Consequences of World War II in the United States*
Rosemary A. Stevens, Charles E. Rosenberg, and Lawton R. Burns, eds., *History and Health Policy in the United States: Putting the Past Back In*
Marianne Sullivan, *Tainted Earth: Smelters, Public Health, and the Environment*
Courtney E. Thompson, *An Organ of Murder: Crime, Violence, and Phrenology in Nineteenth-Century America*

Christine Vatovec, *Dying Green: A Journey through End-of-Life Medicine in Search of Sustainable Health Care*
Barbra Mann Wall, *American Catholic Hospitals: A Century of Changing Markets and Missions*
Frances Ward, *The Door of Last Resort: Memoirs of a Nurse Practitioner*
Jean C. Whelan, *Nursing the Nation: Building the Nurse Labor Force*
Shannon Withycombe, *Lost: Miscarriage in Nineteenth-Century America*

Printed in the United States
by Baker & Taylor Publisher Services